MOBILE CRANE DAILY INSPECTION CHECKLIST BOOK

Company Name: _____

Dates Used:_____

Notes:

Date _____

Operator	
Crane number	

Signature	
Model	

Visual Inspection	Pass	Fail	N/A
Engine fluid level correct (check dip stick or sight glass)			
Hydraulic fluid level correct (check dip stick or sight glass)			
Hydraulic system exhibits no apparent weeping or leaks			
Air system exhibits no audible leaks			
Tire pressure acceptable and tire not damaged			
Telescoping boom exhibits no damage to structure, wear pads, boom stops, or cylinder			
Wire rope free of dirt, excess lube, kinks, and wires and spooled correctly			
Reeving correct			
Wedge sockets and wire rope clips not distorted, cracked, or missing			
Block not damaged			
Ball and hook is free to swivel and rotate			
Guards are in place			
Outrigger float(s) secured with pad pin			
Cab			
Handrails in place and not damaged			
Operator's manual in vehicle			
Load chart legible and visible to operator			
Hand signal chart visible to workers			
Charged fire extinguisher in place			
Cab glass not cracked and wipers are functional			

	Pass	Fail	N/A
Gauges and Indicators			
Load moment indicator operational			
Drum rotation indicator functioning			
Boom length indicator functioning			
Boom angle indicator functioning			
Engine: hydraulic, air, electrical, oil pressure, temperature, and fuel			
Operational Inspection			
Correct counterweight for the load			
Main hoist control functioning			
Auxiliary hoist control functioning			
Anti-two block in place and functioning			
Swing brake			
Lights and horns functional			

Notes:

Date_____

Operator		Signature	
Crane number		Model	

Visual Inspection	Pass	Fail	N/A
Engine fluid level correct (check dip stick or sight glass)			
Hydraulic fluid level correct (check dip stick or sight glass)			
Hydraulic system exhibits no apparent weeping or leaks			
Air system exhibits no audible leaks			
Tire pressure acceptable and tire not damaged			
Telescoping boom exhibits no damage to structure, wear pads, boom stops, or cylinder			
Wire rope free of dirt, excess lube, kinks, and wires and spooled correctly			
Reeving correct			
Wedge sockets and wire rope clips not distorted, cracked, or missing			
Block not damaged			
Ball and hook is free to swivel and rotate			
Guards are in place			
Outrigger float(s) secured with pad pin			
Cab			
Handrails in place and not damaged			
Operator's manual in vehicle			
Load chart legible and visible to operator			
Hand signal chart visible to workers			
Charged fire extinguisher in place			
Cab glass not cracked and wipers are functional			

	Pass	Fail	N/A
Gauges and Indicators			
Load moment indicator operational			
Drum rotation indicator functioning			
Boom length indicator functioning			
Boom angle indicator functioning			
Engine: hydraulic, air, electrical, oil pressure, temperature, and fuel			
Operational Inspection			
Correct counterweight for the load			
Main hoist control functioning			
Auxiliary hoist control functioning			
Anti-two block in place and functioning			
Swing brake			
Lights and horns functional			

Notes:

Date_____

Operator
Crane number

Signature	
Model	

Visual Inspection	Pass	Fail	N/A
Engine fluid level correct (check dip stick or sight glass)			
Hydraulic fluid level correct (check dip stick or sight glass)			
Hydraulic system exhibits no apparent weeping or leaks			
Air system exhibits no audible leaks			
Tire pressure acceptable and tire not damaged			
Telescoping boom exhibits no damage to structure, wear pads, boom stops, or cylinder			
Wire rope free of dirt, excess lube, kinks, and wires and spooled correctly			
Reeving correct			
Wedge sockets and wire rope clips not distorted, cracked, or missing			
Block not damaged			
Ball and hook is free to swivel and rotate			
Guards are in place			
Outrigger float(s) secured with pad pin			
Cab			
Handrails in place and not damaged			
Operator's manual in vehicle			
Load chart legible and visible to operator			
Hand signal chart visible to workers			
Charged fire extinguisher in place			
Cab glass not cracked and wipers are functional			

	Pass	Fail	N/A
Gauges and Indicators			
Load moment indicator operational			
Drum rotation indicator functioning			
Boom length indicator functioning			
Boom angle indicator functioning			
Engine: hydraulic, air, electrical, oil pressure, temperature, and fuel			
Operational Inspection			
Correct counterweight for the load			
Main hoist control functioning			
Auxiliary hoist control functioning			
Anti-two block in place and functioning			
Swing brake			
Lights and horns functional			

Notes:

Date_____

Operator		Signature	
Crane number		Model	

Visual Inspection	Pass	Fail	N/A		Pass	Fail	N/A
Engine fluid level correct (check dip stick or sight glass)				**Gauges and Indicators**			
Hydraulic fluid level correct (check dip stick or sight glass)				Load moment indicator operational			
Hydraulic system exhibits no apparent weeping or leaks				Drum rotation indicator functioning			
Air system exhibits no audible leaks				Boom length indicator functioning			
Tire pressure acceptable and tire not damaged				Boom angle indicator functioning			
Telescoping boom exhibits no damage to structure, wear pads, boom stops, or cylinder				Engine: hydraulic, air, electrical, oil pressure, temperature, and fuel			
Wire rope free of dirt, excess lube, kinks, and wires and spooled correctly				**Operational Inspection**			
Reeving correct				Correct counterweight for the load			
Wedge sockets and wire rope clips not distorted, cracked, or missing				Main hoist control functioning			
Block not damaged				Auxiliary hoist control functioning			
Ball and hook is free to swivel and rotate				Anti-two block in place and functioning			
Guards are in place				Swing brake			
Outrigger float(s) secured with pad pin				Lights and horns functional			
Cab							
Handrails in place and not damaged							
Operator's manual in vehicle							
Load chart legible and visible to operator							
Hand signal chart visible to workers							
Charged fire extinguisher in place							
Cab glass not cracked and wipers are functional							

Notes:

Date_____

Operator
Crane number

Signature	
Model	

Visual Inspection	Pass	Fail	N/A
Engine fluid level correct (check dip stick or sight glass)			
Hydraulic fluid level correct (check dip stick or sight glass)			
Hydraulic system exhibits no apparent weeping or leaks			
Air system exhibits no audible leaks			
Tire pressure acceptable and tire not damaged			
Telescoping boom exhibits no damage to structure, wear pads, boom stops, or cylinder			
Wire rope free of dirt, excess lube, kinks, and wires and spooled correctly			
Reeving correct			
Wedge sockets and wire rope clips not distorted, cracked, or missing			
Block not damaged			
Ball and hook is free to swivel and rotate			
Guards are in place			
Outrigger float(s) secured with pad pin			
Cab			
Handrails in place and not damaged			
Operator's manual in vehicle			
Load chart legible and visible to operator			
Hand signal chart visible to workers			
Charged fire extinguisher in place			
Cab glass not cracked and wipers are functional			

	Pass	Fail	N/A
Gauges and Indicators			
Load moment indicator operational			
Drum rotation indicator functioning			
Boom length indicator functioning			
Boom angle indicator functioning			
Engine: hydraulic, air, electrical, oil pressure, temperature, and fuel			
Operational Inspection			
Correct counterweight for the load			
Main hoist control functioning			
Auxiliary hoist control functioning			
Anti-two block in place and functioning			
Swing brake			
Lights and horns functional			

Notes:

Date_____

Operator		Signature	
Crane number		Model	

Visual Inspection	Pass	Fail	N/A		Pass	Fail	N/A
Engine fluid level correct (check dip stick or sight glass)				**Gauges and Indicators**			
Hydraulic fluid level correct (check dip stick or sight glass)				Load moment indicator operational			
Hydraulic system exhibits no apparent weeping or leaks				Drum rotation indicator functioning			
Air system exhibits no audible leaks				Boom length indicator functioning			
Tire pressure acceptable and tire not damaged				Boom angle indicator functioning			
Telescoping boom exhibits no damage to structure, wear pads, boom stops, or cylinder				Engine: hydraulic, air, electrical, oil pressure, temperature, and fuel			
Wire rope free of dirt, excess lube, kinks, and wires and spooled correctly				**Operational Inspection**			
Reeving correct				Correct counterweight for the load			
Wedge sockets and wire rope clips not distorted, cracked, or missing				Main hoist control functioning			
Block not damaged				Auxiliary hoist control functioning			
Ball and hook is free to swivel and rotate				Anti-two block in place and functioning			
Guards are in place				Swing brake			
Outrigger float(s) secured with pad pin				Lights and horns functional			
Cab							
Handrails in place and not damaged							
Operator's manual in vehicle							
Load chart legible and visible to operator							
Hand signal chart visible to workers							
Charged fire extinguisher in place							
Cab glass not cracked and wipers are functional							

Notes:

Date_____

Operator
Crane number

Signature	
Model	

Visual Inspection	Pass	Fail	N/A
Engine fluid level correct (check dip stick or sight glass)			
Hydraulic fluid level correct (check dip stick or sight glass)			
Hydraulic system exhibits no apparent weeping or leaks			
Air system exhibits no audible leaks			
Tire pressure acceptable and tire not damaged			
Telescoping boom exhibits no damage to structure, wear pads, boom stops, or cylinder			
Wire rope free of dirt, excess lube, kinks, and wires and spooled correctly			
Reeving correct			
Wedge sockets and wire rope clips not distorted, cracked, or missing			
Block not damaged			
Ball and hook is free to swivel and rotate			
Guards are in place			
Outrigger float(s) secured with pad pin			
Cab			
Handrails in place and not damaged			
Operator's manual in vehicle			
Load chart legible and visible to operator			
Hand signal chart visible to workers			
Charged fire extinguisher in place			
Cab glass not cracked and wipers are functional			

	Pass	Fail	N/A
Gauges and Indicators			
Load moment indicator operational			
Drum rotation indicator functioning			
Boom length indicator functioning			
Boom angle indicator functioning			
Engine: hydraulic, air, electrical, oil pressure, temperature, and fuel			
Operational Inspection			
Correct counterweight for the load			
Main hoist control functioning			
Auxiliary hoist control functioning			
Anti-two block in place and functioning			
Swing brake			
Lights and horns functional			

Notes:

Date_____

Operator		Signature	
Crane number		Model	

Visual Inspection	Pass	Fail	N/A
Engine fluid level correct (check dip stick or sight glass)			
Hydraulic fluid level correct (check dip stick or sight glass)			
Hydraulic system exhibits no apparent weeping or leaks			
Air system exhibits no audible leaks			
Tire pressure acceptable and tire not damaged			
Telescoping boom exhibits no damage to structure, wear pads, boom stops, or cylinder			
Wire rope free of dirt, excess lube, kinks, and wires and spooled correctly			
Reeving correct			
Wedge sockets and wire rope clips not distorted, cracked, or missing			
Block not damaged			
Ball and hook is free to swivel and rotate			
Guards are in place			
Outrigger float(s) secured with pad pin			
Cab			
Handrails in place and not damaged			
Operator's manual in vehicle			
Load chart legible and visible to operator			
Hand signal chart visible to workers			
Charged fire extinguisher in place			
Cab glass not cracked and wipers are functional			

	Pass	Fail	N/A
Gauges and Indicators			
Load moment indicator operational			
Drum rotation indicator functioning			
Boom length indicator functioning			
Boom angle indicator functioning			
Engine: hydraulic, air, electrical, oil pressure, temperature, and fuel			
Operational Inspection			
Correct counterweight for the load			
Main hoist control functioning			
Auxiliary hoist control functioning			
Anti-two block in place and functioning			
Swing brake			
Lights and horns functional			

Notes:

Date_____

Operator
Crane number

Signature	
Model	

Visual Inspection	Pass	Fail	N/A
Engine fluid level correct (check dip stick or sight glass)			
Hydraulic fluid level correct (check dip stick or sight glass)			
Hydraulic system exhibits no apparent weeping or leaks			
Air system exhibits no audible leaks			
Tire pressure acceptable and tire not damaged			
Telescoping boom exhibits no damage to structure, wear pads, boom stops, or cylinder			
Wire rope free of dirt, excess lube, kinks, and wires and spooled correctly			
Reeving correct			
Wedge sockets and wire rope clips not distorted, cracked, or missing			
Block not damaged			
Ball and hook is free to swivel and rotate			
Guards are in place			
Outrigger float(s) secured with pad pin			
Cab			
Handrails in place and not damaged			
Operator's manual in vehicle			
Load chart legible and visible to operator			
Hand signal chart visible to workers			
Charged fire extinguisher in place			
Cab glass not cracked and wipers are functional			

	Pass	Fail	N/A
Gauges and Indicators			
Load moment indicator operational			
Drum rotation indicator functioning			
Boom length indicator functioning			
Boom angle indicator functioning			
Engine: hydraulic, air, electrical, oil pressure, temperature, and fuel			
Operational Inspection			
Correct counterweight for the load			
Main hoist control functioning			
Auxiliary hoist control functioning			
Anti-two block in place and functioning			
Swing brake			
Lights and horns functional			

Notes:

Date_____

Operator
Crane number

Signature	
Model	

Visual Inspection	Pass	Fail	N/A
Engine fluid level correct (check dip stick or sight glass)			
Hydraulic fluid level correct (check dip stick or sight glass)			
Hydraulic system exhibits no apparent weeping or leaks			
Air system exhibits no audible leaks			
Tire pressure acceptable and tire not damaged			
Telescoping boom exhibits no damage to structure, wear pads, boom stops, or cylinder			
Wire rope free of dirt, excess lube, kinks, and wires and spooled correctly			
Reeving correct			
Wedge sockets and wire rope clips not distorted, cracked, or missing			
Block not damaged			
Ball and hook is free to swivel and rotate			
Guards are in place			
Outrigger float(s) secured with pad pin			
Cab			
Handrails in place and not damaged			
Operator's manual in vehicle			
Load chart legible and visible to operator			
Hand signal chart visible to workers			
Charged fire extinguisher in place			
Cab glass not cracked and wipers are functional			

	Pass	Fail	N/A
Gauges and Indicators			
Load moment indicator operational			
Drum rotation indicator functioning			
Boom length indicator functioning			
Boom angle indicator functioning			
Engine: hydraulic, air, electrical, oil pressure, temperature, and fuel			
Operational Inspection			
Correct counterweight for the load			
Main hoist control functioning			
Auxiliary hoist control functioning			
Anti-two block in place and functioning			
Swing brake			
Lights and horns functional			

Notes:

Date_____

Operator
Crane number

Signature	
Model	

Visual Inspection	Pass	Fail	N/A
Engine fluid level correct (check dip stick or sight glass)			
Hydraulic fluid level correct (check dip stick or sight glass)			
Hydraulic system exhibits no apparent weeping or leaks			
Air system exhibits no audible leaks			
Tire pressure acceptable and tire not damaged			
Telescoping boom exhibits no damage to structure, wear pads, boom stops, or cylinder			
Wire rope free of dirt, excess lube, kinks, and wires and spooled correctly			
Reeving correct			
Wedge sockets and wire rope clips not distorted, cracked, or missing			
Block not damaged			
Ball and hook is free to swivel and rotate			
Guards are in place			
Outrigger float(s) secured with pad pin			
Cab			
Handrails in place and not damaged			
Operator's manual in vehicle			
Load chart legible and visible to operator			
Hand signal chart visible to workers			
Charged fire extinguisher in place			
Cab glass not cracked and wipers are functional			

	Pass	Fail	N/A
Gauges and Indicators			
Load moment indicator operational			
Drum rotation indicator functioning			
Boom length indicator functioning			
Boom angle indicator functioning			
Engine: hydraulic, air, electrical, oil pressure, temperature, and fuel			
Operational Inspection			
Correct counterweight for the load			
Main hoist control functioning			
Auxiliary hoist control functioning			
Anti-two block in place and functioning			
Swing brake			
Lights and horns functional			

Notes:

Date_____

Operator		Signature	
Crane number		Model	

Visual Inspection	Pass	Fail	N/A		Pass	Fail	N/A
Engine fluid level correct (check dip stick or sight glass)				**Gauges and Indicators**			
Hydraulic fluid level correct (check dip stick or sight glass)				Load moment indicator operational			
Hydraulic system exhibits no apparent weeping or leaks				Drum rotation indicator functioning			
Air system exhibits no audible leaks				Boom length indicator functioning			
Tire pressure acceptable and tire not damaged				Boom angle indicator functioning			
Telescoping boom exhibits no damage to structure, wear pads, boom stops, or cylinder				Engine: hydraulic, air, electrical, oil pressure, temperature, and fuel			
Wire rope free of dirt, excess lube, kinks, and wires and spooled correctly				**Operational Inspection**			
Reeving correct				Correct counterweight for the load			
Wedge sockets and wire rope clips not distorted, cracked, or missing				Main hoist control functioning			
Block not damaged				Auxiliary hoist control functioning			
Ball and hook is free to swivel and rotate				Anti-two block in place and functioning			
Guards are in place				Swing brake			
Outrigger float(s) secured with pad pin				Lights and horns functional			
Cab							
Handrails in place and not damaged							
Operator's manual in vehicle							
Load chart legible and visible to operator							
Hand signal chart visible to workers							
Charged fire extinguisher in place							
Cab glass not cracked and wipers are functional							

Notes:

Date_____

Operator
Crane number

Signature	
Model	

Visual Inspection	Pass	Fail	N/A
Engine fluid level correct (check dip stick or sight glass)			
Hydraulic fluid level correct (check dip stick or sight glass)			
Hydraulic system exhibits no apparent weeping or leaks			
Air system exhibits no audible leaks			
Tire pressure acceptable and tire not damaged			
Telescoping boom exhibits no damage to structure, wear pads, boom stops, or cylinder			
Wire rope free of dirt, excess lube, kinks, and wires and spooled correctly			
Reeving correct			
Wedge sockets and wire rope clips not distorted, cracked, or missing			
Block not damaged			
Ball and hook is free to swivel and rotate			
Guards are in place			
Outrigger float(s) secured with pad pin			
Cab			
Handrails in place and not damaged			
Operator's manual in vehicle			
Load chart legible and visible to operator			
Hand signal chart visible to workers			
Charged fire extinguisher in place			
Cab glass not cracked and wipers are functional			

	Pass	Fail	N/A
Gauges and Indicators			
Load moment indicator operational			
Drum rotation indicator functioning			
Boom length indicator functioning			
Boom angle indicator functioning			
Engine: hydraulic, air, electrical, oil pressure, temperature, and fuel			
Operational Inspection			
Correct counterweight for the load			
Main hoist control functioning			
Auxiliary hoist control functioning			
Anti-two block in place and functioning			
Swing brake			
Lights and horns functional			

Notes:

Date_____

Operator		Signature	
Crane number		Model	

Visual Inspection	Pass	Fail	N/A
Engine fluid level correct (check dip stick or sight glass)			
Hydraulic fluid level correct (check dip stick or sight glass)			
Hydraulic system exhibits no apparent weeping or leaks			
Air system exhibits no audible leaks			
Tire pressure acceptable and tire not damaged			
Telescoping boom exhibits no damage to structure, wear pads, boom stops, or cylinder			
Wire rope free of dirt, excess lube, kinks, and wires and spooled correctly			
Reeving correct			
Wedge sockets and wire rope clips not distorted, cracked, or missing			
Block not damaged			
Ball and hook is free to swivel and rotate			
Guards are in place			
Outrigger float(s) secured with pad pin			
Cab			
Handrails in place and not damaged			
Operator's manual in vehicle			
Load chart legible and visible to operator			
Hand signal chart visible to workers			
Charged fire extinguisher in place			
Cab glass not cracked and wipers are functional			

	Pass	Fail	N/A
Gauges and Indicators			
Load moment indicator operational			
Drum rotation indicator functioning			
Boom length indicator functioning			
Boom angle indicator functioning			
Engine: hydraulic, air, electrical, oil pressure, temperature, and fuel			
Operational Inspection			
Correct counterweight for the load			
Main hoist control functioning			
Auxiliary hoist control functioning			
Anti-two block in place and functioning			
Swing brake			
Lights and horns functional			

Notes:

Date_____

Operator	
Crane number	

Signature	
Model	

Visual Inspection	Pass	Fail	N/A
Engine fluid level correct (check dip stick or sight glass)			
Hydraulic fluid level correct (check dip stick or sight glass)			
Hydraulic system exhibits no apparent weeping or leaks			
Air system exhibits no audible leaks			
Tire pressure acceptable and tire not damaged			
Telescoping boom exhibits no damage to structure, wear pads, boom stops, or cylinder			
Wire rope free of dirt, excess lube, kinks, and wires and spooled correctly			
Reeving correct			
Wedge sockets and wire rope clips not distorted, cracked, or missing			
Block not damaged			
Ball and hook is free to swivel and rotate			
Guards are in place			
Outrigger float(s) secured with pad pin			
Cab			
Handrails in place and not damaged			
Operator's manual in vehicle			
Load chart legible and visible to operator			
Hand signal chart visible to workers			
Charged fire extinguisher in place			
Cab glass not cracked and wipers are functional			

	Pass	Fail	N/A
Gauges and Indicators			
Load moment indicator operational			
Drum rotation indicator functioning			
Boom length indicator functioning			
Boom angle indicator functioning			
Engine: hydraulic, air, electrical, oil pressure, temperature, and fuel			
Operational Inspection			
Correct counterweight for the load			
Main hoist control functioning			
Auxiliary hoist control functioning			
Anti-two block in place and functioning			
Swing brake			
Lights and horns functional			

Notes:

Date_____

Operator		Signature	
Crane number		Model	

Visual Inspection	Pass	Fail	N/A
Engine fluid level correct (check dip stick or sight glass)			
Hydraulic fluid level correct (check dip stick or sight glass)			
Hydraulic system exhibits no apparent weeping or leaks			
Air system exhibits no audible leaks			
Tire pressure acceptable and tire not damaged			
Telescoping boom exhibits no damage to structure, wear pads, boom stops, or cylinder			
Wire rope free of dirt, excess lube, kinks, and wires and spooled correctly			
Reeving correct			
Wedge sockets and wire rope clips not distorted, cracked, or missing			
Block not damaged			
Ball and hook is free to swivel and rotate			
Guards are in place			
Outrigger float(s) secured with pad pin			
Cab			
Handrails in place and not damaged			
Operator's manual in vehicle			
Load chart legible and visible to operator			
Hand signal chart visible to workers			
Charged fire extinguisher in place			
Cab glass not cracked and wipers are functional			

	Pass	Fail	N/A
Gauges and Indicators			
Load moment indicator operational			
Drum rotation indicator functioning			
Boom length indicator functioning			
Boom angle indicator functioning			
Engine: hydraulic, air, electrical, oil pressure, temperature, and fuel			
Operational Inspection			
Correct counterweight for the load			
Main hoist control functioning			
Auxiliary hoist control functioning			
Anti-two block in place and functioning			
Swing brake			
Lights and horns functional			

Notes:

Date_____

Operator
Crane number

Signature	
Model	

Visual Inspection	Pass	Fail	N/A
Engine fluid level correct (check dip stick or sight glass)			
Hydraulic fluid level correct (check dip stick or sight glass)			
Hydraulic system exhibits no apparent weeping or leaks			
Air system exhibits no audible leaks			
Tire pressure acceptable and tire not damaged			
Telescoping boom exhibits no damage to structure, wear pads, boom stops, or cylinder			
Wire rope free of dirt, excess lube, kinks, and wires and spooled correctly			
Reeving correct			
Wedge sockets and wire rope clips not distorted, cracked, or missing			
Block not damaged			
Ball and hook is free to swivel and rotate			
Guards are in place			
Outrigger float(s) secured with pad pin			
Cab			
Handrails in place and not damaged			
Operator's manual in vehicle			
Load chart legible and visible to operator			
Hand signal chart visible to workers			
Charged fire extinguisher in place			
Cab glass not cracked and wipers are functional			

	Pass	Fail	N/A
Gauges and Indicators			
Load moment indicator operational			
Drum rotation indicator functioning			
Boom length indicator functioning			
Boom angle indicator functioning			
Engine: hydraulic, air, electrical, oil pressure, temperature, and fuel			
Operational Inspection			
Correct counterweight for the load			
Main hoist control functioning			
Auxiliary hoist control functioning			
Anti-two block in place and functioning			
Swing brake			
Lights and horns functional			

Notes:

Date_____

Operator		Signature	
Crane number		Model	

Visual Inspection	Pass	Fail	N/A		Pass	Fail	N/A
Engine fluid level correct (check dip stick or sight glass)				**Gauges and Indicators**			
Hydraulic fluid level correct (check dip stick or sight glass)				Load moment indicator operational			
Hydraulic system exhibits no apparent weeping or leaks				Drum rotation indicator functioning			
Air system exhibits no audible leaks				Boom length indicator functioning			
Tire pressure acceptable and tire not damaged				Boom angle indicator functioning			
Telescoping boom exhibits no damage to structure, wear pads, boom stops, or cylinder				Engine: hydraulic, air, electrical, oil pressure, temperature, and fuel			
Wire rope free of dirt, excess lube, kinks, and wires and spooled correctly				**Operational Inspection**			
Reeving correct				Correct counterweight for the load			
Wedge sockets and wire rope clips not distorted, cracked, or missing				Main hoist control functioning			
Block not damaged				Auxiliary hoist control functioning			
Ball and hook is free to swivel and rotate				Anti-two block in place and functioning			
Guards are in place				Swing brake			
Outrigger float(s) secured with pad pin				Lights and horns functional			
Cab							
Handrails in place and not damaged							
Operator's manual in vehicle							
Load chart legible and visible to operator							
Hand signal chart visible to workers							
Charged fire extinguisher in place							
Cab glass not cracked and wipers are functional							

Notes:

Date_____

Operator
Crane number

Signature	
Model	

Visual Inspection	Pass	Fail	N/A
Engine fluid level correct (check dip stick or sight glass)			
Hydraulic fluid level correct (check dip stick or sight glass)			
Hydraulic system exhibits no apparent weeping or leaks			
Air system exhibits no audible leaks			
Tire pressure acceptable and tire not damaged			
Telescoping boom exhibits no damage to structure, wear pads, boom stops, or cylinder			
Wire rope free of dirt, excess lube, kinks, and wires and spooled correctly			
Reeving correct			
Wedge sockets and wire rope clips not distorted, cracked, or missing			
Block not damaged			
Ball and hook is free to swivel and rotate			
Guards are in place			
Outrigger float(s) secured with pad pin			
Cab			
Handrails in place and not damaged			
Operator's manual in vehicle			
Load chart legible and visible to operator			
Hand signal chart visible to workers			
Charged fire extinguisher in place			
Cab glass not cracked and wipers are functional			

	Pass	Fail	N/A
Gauges and Indicators			
Load moment indicator operational			
Drum rotation indicator functioning			
Boom length indicator functioning			
Boom angle indicator functioning			
Engine: hydraulic, air, electrical, oil pressure, temperature, and fuel			
Operational Inspection			
Correct counterweight for the load			
Main hoist control functioning			
Auxiliary hoist control functioning			
Anti-two block in place and functioning			
Swing brake			
Lights and horns functional			

Notes:

Date_____

Operator		Signature	
Crane number		Model	

Visual Inspection	Pass	Fail	N/A		Pass	Fail	N/A
Engine fluid level correct (check dip stick or sight glass)				**Gauges and Indicators**			
Hydraulic fluid level correct (check dip stick or sight glass)				Load moment indicator operational			
Hydraulic system exhibits no apparent weeping or leaks				Drum rotation indicator functioning			
Air system exhibits no audible leaks				Boom length indicator functioning			
Tire pressure acceptable and tire not damaged				Boom angle indicator functioning			
Telescoping boom exhibits no damage to structure, wear pads, boom stops, or cylinder				Engine: hydraulic, air, electrical, oil pressure, temperature, and fuel			
Wire rope free of dirt, excess lube, kinks, and wires and spooled correctly				**Operational Inspection**			
Reeving correct				Correct counterweight for the load			
Wedge sockets and wire rope clips not distorted, cracked, or missing				Main hoist control functioning			
Block not damaged				Auxiliary hoist control functioning			
Ball and hook is free to swivel and rotate				Anti-two block in place and functioning			
Guards are in place				Swing brake			
Outrigger float(s) secured with pad pin				Lights and horns functional			
Cab							
Handrails in place and not damaged							
Operator's manual in vehicle							
Load chart legible and visible to operator							
Hand signal chart visible to workers							
Charged fire extinguisher in place							
Cab glass not cracked and wipers are functional							

Notes:

Date_____

Operator
Crane number

Signature	
Model	

Visual Inspection	Pass	Fail	N/A
Engine fluid level correct (check dip stick or sight glass)			
Hydraulic fluid level correct (check dip stick or sight glass)			
Hydraulic system exhibits no apparent weeping or leaks			
Air system exhibits no audible leaks			
Tire pressure acceptable and tire not damaged			
Telescoping boom exhibits no damage to structure, wear pads, boom stops, or cylinder			
Wire rope free of dirt, excess lube, kinks, and wires and spooled correctly			
Reeving correct			
Wedge sockets and wire rope clips not distorted, cracked, or missing			
Block not damaged			
Ball and hook is free to swivel and rotate			
Guards are in place			
Outrigger float(s) secured with pad pin			
Cab			
Handrails in place and not damaged			
Operator's manual in vehicle			
Load chart legible and visible to operator			
Hand signal chart visible to workers			
Charged fire extinguisher in place			
Cab glass not cracked and wipers are functional			

	Pass	Fail	N/A
Gauges and Indicators			
Load moment indicator operational			
Drum rotation indicator functioning			
Boom length indicator functioning			
Boom angle indicator functioning			
Engine: hydraulic, air, electrical, oil pressure, temperature, and fuel			
Operational Inspection			
Correct counterweight for the load			
Main hoist control functioning			
Auxiliary hoist control functioning			
Anti-two block in place and functioning			
Swing brake			
Lights and horns functional			

Notes:

Date_____

Operator		Signature	
Crane number		Model	

Visual Inspection	Pass	Fail	N/A		Pass	Fail	N/A
Engine fluid level correct (check dip stick or sight glass)				**Gauges and Indicators**			
Hydraulic fluid level correct (check dip stick or sight glass)				Load moment indicator operational			
Hydraulic system exhibits no apparent weeping or leaks				Drum rotation indicator functioning			
Air system exhibits no audible leaks				Boom length indicator functioning			
Tire pressure acceptable and tire not damaged				Boom angle indicator functioning			
Telescoping boom exhibits no damage to structure, wear pads, boom stops, or cylinder				Engine: hydraulic, air, electrical, oil pressure, temperature, and fuel			
Wire rope free of dirt, excess lube, kinks, and wires and spooled correctly				**Operational Inspection**			
Reeving correct				Correct counterweight for the load			
Wedge sockets and wire rope clips not distorted, cracked, or missing				Main hoist control functioning			
Block not damaged				Auxiliary hoist control functioning			
Ball and hook is free to swivel and rotate				Anti-two block in place and functioning			
Guards are in place				Swing brake			
Outrigger float(s) secured with pad pin				Lights and horns functional			
Cab							
Handrails in place and not damaged							
Operator's manual in vehicle							
Load chart legible and visible to operator							
Hand signal chart visible to workers							
Charged fire extinguisher in place							
Cab glass not cracked and wipers are functional							

Notes:

Date_____

Operator
Crane number

Signature	
Model	

Visual Inspection	Pass	Fail	N/A
Engine fluid level correct (check dip stick or sight glass)			
Hydraulic fluid level correct (check dip stick or sight glass)			
Hydraulic system exhibits no apparent weeping or leaks			
Air system exhibits no audible leaks			
Tire pressure acceptable and tire not damaged			
Telescoping boom exhibits no damage to structure, wear pads, boom stops, or cylinder			
Wire rope free of dirt, excess lube, kinks, and wires and spooled correctly			
Reeving correct			
Wedge sockets and wire rope clips not distorted, cracked, or missing			
Block not damaged			
Ball and hook is free to swivel and rotate			
Guards are in place			
Outrigger float(s) secured with pad pin			
Cab			
Handrails in place and not damaged			
Operator's manual in vehicle			
Load chart legible and visible to operator			
Hand signal chart visible to workers			
Charged fire extinguisher in place			
Cab glass not cracked and wipers are functional			

	Pass	Fail	N/A
Gauges and Indicators			
Load moment indicator operational			
Drum rotation indicator functioning			
Boom length indicator functioning			
Boom angle indicator functioning			
Engine: hydraulic, air, electrical, oil pressure, temperature, and fuel			
Operational Inspection			
Correct counterweight for the load			
Main hoist control functioning			
Auxiliary hoist control functioning			
Anti-two block in place and functioning			
Swing brake			
Lights and horns functional			

Notes:

Date_____

Operator		Signature	
Crane number		Model	

Visual Inspection	Pass	Fail	N/A
Engine fluid level correct (check dip stick or sight glass)			
Hydraulic fluid level correct (check dip stick or sight glass)			
Hydraulic system exhibits no apparent weeping or leaks			
Air system exhibits no audible leaks			
Tire pressure acceptable and tire not damaged			
Telescoping boom exhibits no damage to structure, wear pads, boom stops, or cylinder			
Wire rope free of dirt, excess lube, kinks, and wires and spooled correctly			
Reeving correct			
Wedge sockets and wire rope clips not distorted, cracked, or missing			
Block not damaged			
Ball and hook is free to swivel and rotate			
Guards are in place			
Outrigger float(s) secured with pad pin			
Cab			
Handrails in place and not damaged			
Operator's manual in vehicle			
Load chart legible and visible to operator			
Hand signal chart visible to workers			
Charged fire extinguisher in place			
Cab glass not cracked and wipers are functional			

	Pass	Fail	N/A
Gauges and Indicators			
Load moment indicator operational			
Drum rotation indicator functioning			
Boom length indicator functioning			
Boom angle indicator functioning			
Engine: hydraulic, air, electrical, oil pressure, temperature, and fuel			
Operational Inspection			
Correct counterweight for the load			
Main hoist control functioning			
Auxiliary hoist control functioning			
Anti-two block in place and functioning			
Swing brake			
Lights and horns functional			

Notes:

Date_____

Operator
Crane number

Signature	
Model	

Visual Inspection	Pass	Fail	N/A
Engine fluid level correct (check dip stick or sight glass)			
Hydraulic fluid level correct (check dip stick or sight glass)			
Hydraulic system exhibits no apparent weeping or leaks			
Air system exhibits no audible leaks			
Tire pressure acceptable and tire not damaged			
Telescoping boom exhibits no damage to structure, wear pads, boom stops, or cylinder			
Wire rope free of dirt, excess lube, kinks, and wires and spooled correctly			
Reeving correct			
Wedge sockets and wire rope clips not distorted, cracked, or missing			
Block not damaged			
Ball and hook is free to swivel and rotate			
Guards are in place			
Outrigger float(s) secured with pad pin			
Cab			
Handrails in place and not damaged			
Operator's manual in vehicle			
Load chart legible and visible to operator			
Hand signal chart visible to workers			
Charged fire extinguisher in place			
Cab glass not cracked and wipers are functional			

	Pass	Fail	N/A
Gauges and Indicators			
Load moment indicator operational			
Drum rotation indicator functioning			
Boom length indicator functioning			
Boom angle indicator functioning			
Engine: hydraulic, air, electrical, oil pressure, temperature, and fuel			
Operational Inspection			
Correct counterweight for the load			
Main hoist control functioning			
Auxiliary hoist control functioning			
Anti-two block in place and functioning			
Swing brake			
Lights and horns functional			

Notes:

Date_____

Operator
Crane number

Signature	
Model	

Visual Inspection	Pass	Fail	N/A
Engine fluid level correct (check dip stick or sight glass)			
Hydraulic fluid level correct (check dip stick or sight glass)			
Hydraulic system exhibits no apparent weeping or leaks			
Air system exhibits no audible leaks			
Tire pressure acceptable and tire not damaged			
Telescoping boom exhibits no damage to structure, wear pads, boom stops, or cylinder			
Wire rope free of dirt, excess lube, kinks, and wires and spooled correctly			
Reeving correct			
Wedge sockets and wire rope clips not distorted, cracked, or missing			
Block not damaged			
Ball and hook is free to swivel and rotate			
Guards are in place			
Outrigger float(s) secured with pad pin			
Cab			
Handrails in place and not damaged			
Operator's manual in vehicle			
Load chart legible and visible to operator			
Hand signal chart visible to workers			
Charged fire extinguisher in place			
Cab glass not cracked and wipers are functional			

	Pass	Fail	N/A
Gauges and Indicators			
Load moment indicator operational			
Drum rotation indicator functioning			
Boom length indicator functioning			
Boom angle indicator functioning			
Engine: hydraulic, air, electrical, oil pressure, temperature, and fuel			
Operational Inspection			
Correct counterweight for the load			
Main hoist control functioning			
Auxiliary hoist control functioning			
Anti-two block in place and functioning			
Swing brake			
Lights and horns functional			

Notes:

Date_____

Operator
Crane number

Signature	
Model	

Visual Inspection	Pass	Fail	N/A
Engine fluid level correct (check dip stick or sight glass)			
Hydraulic fluid level correct (check dip stick or sight glass)			
Hydraulic system exhibits no apparent weeping or leaks			
Air system exhibits no audible leaks			
Tire pressure acceptable and tire not damaged			
Telescoping boom exhibits no damage to structure, wear pads, boom stops, or cylinder			
Wire rope free of dirt, excess lube, kinks, and wires and spooled correctly			
Reeving correct			
Wedge sockets and wire rope clips not distorted, cracked, or missing			
Block not damaged			
Ball and hook is free to swivel and rotate			
Guards are in place			
Outrigger float(s) secured with pad pin			
Cab			
Handrails in place and not damaged			
Operator's manual in vehicle			
Load chart legible and visible to operator			
Hand signal chart visible to workers			
Charged fire extinguisher in place			
Cab glass not cracked and wipers are functional			

	Pass	Fail	N/A
Gauges and Indicators			
Load moment indicator operational			
Drum rotation indicator functioning			
Boom length indicator functioning			
Boom angle indicator functioning			
Engine: hydraulic, air, electrical, oil pressure, temperature, and fuel			
Operational Inspection			
Correct counterweight for the load			
Main hoist control functioning			
Auxiliary hoist control functioning			
Anti-two block in place and functioning			
Swing brake			
Lights and horns functional			

Notes:

Date_____

Operator		Signature	
Crane number		Model	

Visual Inspection	Pass	Fail	N/A		Pass	Fail	N/A
Engine fluid level correct (check dip stick or sight glass)				**Gauges and Indicators**			
Hydraulic fluid level correct (check dip stick or sight glass)				Load moment indicator operational			
Hydraulic system exhibits no apparent weeping or leaks				Drum rotation indicator functioning			
Air system exhibits no audible leaks				Boom length indicator functioning			
Tire pressure acceptable and tire not damaged				Boom angle indicator functioning			
Telescoping boom exhibits no damage to structure, wear pads, boom stops, or cylinder				Engine: hydraulic, air, electrical, oil pressure, temperature, and fuel			
Wire rope free of dirt, excess lube, kinks, and wires and spooled correctly				**Operational Inspection**			
Reeving correct				Correct counterweight for the load			
Wedge sockets and wire rope clips not distorted, cracked, or missing				Main hoist control functioning			
Block not damaged				Auxiliary hoist control functioning			
Ball and hook is free to swivel and rotate				Anti-two block in place and functioning			
Guards are in place				Swing brake			
Outrigger float(s) secured with pad pin				Lights and horns functional			
Cab							
Handrails in place and not damaged							
Operator's manual in vehicle							
Load chart legible and visible to operator							
Hand signal chart visible to workers							
Charged fire extinguisher in place							
Cab glass not cracked and wipers are functional							

Notes:

Date_____

Operator
Crane number

Signature	
Model	

Visual Inspection	Pass	Fail	N/A
Engine fluid level correct (check dip stick or sight glass)			
Hydraulic fluid level correct (check dip stick or sight glass)			
Hydraulic system exhibits no apparent weeping or leaks			
Air system exhibits no audible leaks			
Tire pressure acceptable and tire not damaged			
Telescoping boom exhibits no damage to structure, wear pads, boom stops, or cylinder			
Wire rope free of dirt, excess lube, kinks, and wires and spooled correctly			
Reeving correct			
Wedge sockets and wire rope clips not distorted, cracked, or missing			
Block not damaged			
Ball and hook is free to swivel and rotate			
Guards are in place			
Outrigger float(s) secured with pad pin			
Cab			
Handrails in place and not damaged			
Operator's manual in vehicle			
Load chart legible and visible to operator			
Hand signal chart visible to workers			
Charged fire extinguisher in place			
Cab glass not cracked and wipers are functional			

	Pass	Fail	N/A
Gauges and Indicators			
Load moment indicator operational			
Drum rotation indicator functioning			
Boom length indicator functioning			
Boom angle indicator functioning			
Engine: hydraulic, air, electrical, oil pressure, temperature, and fuel			
Operational Inspection			
Correct counterweight for the load			
Main hoist control functioning			
Auxiliary hoist control functioning			
Anti-two block in place and functioning			
Swing brake			
Lights and horns functional			

Notes:

Date_____

Operator		Signature	
Crane number		Model	

Visual Inspection	Pass	Fail	N/A		Pass	Fail	N/A
Engine fluid level correct (check dip stick or sight glass)				**Gauges and Indicators**			
Hydraulic fluid level correct (check dip stick or sight glass)				Load moment indicator operational			
Hydraulic system exhibits no apparent weeping or leaks				Drum rotation indicator functioning			
Air system exhibits no audible leaks				Boom length indicator functioning			
Tire pressure acceptable and tire not damaged				Boom angle indicator functioning			
Telescoping boom exhibits no damage to structure, wear pads, boom stops, or cylinder				Engine: hydraulic, air, electrical, oil pressure, temperature, and fuel			
Wire rope free of dirt, excess lube, kinks, and wires and spooled correctly				**Operational Inspection**			
Reeving correct				Correct counterweight for the load			
Wedge sockets and wire rope clips not distorted, cracked, or missing				Main hoist control functioning			
Block not damaged				Auxiliary hoist control functioning			
Ball and hook is free to swivel and rotate				Anti-two block in place and functioning			
Guards are in place				Swing brake			
Outrigger float(s) secured with pad pin				Lights and horns functional			
Cab							
Handrails in place and not damaged							
Operator's manual in vehicle							
Load chart legible and visible to operator							
Hand signal chart visible to workers							
Charged fire extinguisher in place							
Cab glass not cracked and wipers are functional							

Notes:

Date _____

Operator		Signature	
Crane number		Model	

Visual Inspection	Pass	Fail	N/A
Engine fluid level correct (check dip stick or sight glass)			
Hydraulic fluid level correct (check dip stick or sight glass)			
Hydraulic system exhibits no apparent weeping or leaks			
Air system exhibits no audible leaks			
Tire pressure acceptable and tire not damaged			
Telescoping boom exhibits no damage to structure, wear pads, boom stops, or cylinder			
Wire rope free of dirt, excess lube, kinks, and wires and spooled correctly			
Reeving correct			
Wedge sockets and wire rope clips not distorted, cracked, or missing			
Block not damaged			
Ball and hook is free to swivel and rotate			
Guards are in place			
Outrigger float(s) secured with pad pin			
Cab			
Handrails in place and not damaged			
Operator's manual in vehicle			
Load chart legible and visible to operator			
Hand signal chart visible to workers			
Charged fire extinguisher in place			
Cab glass not cracked and wipers are functional			

	Pass	Fail	N/A
Gauges and Indicators			
Load moment indicator operational			
Drum rotation indicator functioning			
Boom length indicator functioning			
Boom angle indicator functioning			
Engine: hydraulic, air, electrical, oil pressure, temperature, and fuel			
Operational Inspection			
Correct counterweight for the load			
Main hoist control functioning			
Auxiliary hoist control functioning			
Anti-two block in place and functioning			
Swing brake			
Lights and horns functional			

Notes:

Date_____

Operator		Signature	
Crane number		Model	

Visual Inspection	Pass	Fail	N/A
Engine fluid level correct (check dip stick or sight glass)			
Hydraulic fluid level correct (check dip stick or sight glass)			
Hydraulic system exhibits no apparent weeping or leaks			
Air system exhibits no audible leaks			
Tire pressure acceptable and tire not damaged			
Telescoping boom exhibits no damage to structure, wear pads, boom stops, or cylinder			
Wire rope free of dirt, excess lube, kinks, and wires and spooled correctly			
Reeving correct			
Wedge sockets and wire rope clips not distorted, cracked, or missing			
Block not damaged			
Ball and hook is free to swivel and rotate			
Guards are in place			
Outrigger float(s) secured with pad pin			
Cab			
Handrails in place and not damaged			
Operator's manual in vehicle			
Load chart legible and visible to operator			
Hand signal chart visible to workers			
Charged fire extinguisher in place			
Cab glass not cracked and wipers are functional			

	Pass	Fail	N/A
Gauges and Indicators			
Load moment indicator operational			
Drum rotation indicator functioning			
Boom length indicator functioning			
Boom angle indicator functioning			
Engine: hydraulic, air, electrical, oil pressure, temperature, and fuel			
Operational Inspection			
Correct counterweight for the load			
Main hoist control functioning			
Auxiliary hoist control functioning			
Anti-two block in place and functioning			
Swing brake			
Lights and horns functional			

Notes:

Date_____

Operator
Crane number

Signature	
Model	

Visual Inspection	Pass	Fail	N/A
Engine fluid level correct (check dip stick or sight glass)			
Hydraulic fluid level correct (check dip stick or sight glass)			
Hydraulic system exhibits no apparent weeping or leaks			
Air system exhibits no audible leaks			
Tire pressure acceptable and tire not damaged			
Telescoping boom exhibits no damage to structure, wear pads, boom stops, or cylinder			
Wire rope free of dirt, excess lube, kinks, and wires and spooled correctly			
Reeving correct			
Wedge sockets and wire rope clips not distorted, cracked, or missing			
Block not damaged			
Ball and hook is free to swivel and rotate			
Guards are in place			
Outrigger float(s) secured with pad pin			
Cab			
Handrails in place and not damaged			
Operator's manual in vehicle			
Load chart legible and visible to operator			
Hand signal chart visible to workers			
Charged fire extinguisher in place			
Cab glass not cracked and wipers are functional			

	Pass	Fail	N/A
Gauges and Indicators			
Load moment indicator operational			
Drum rotation indicator functioning			
Boom length indicator functioning			
Boom angle indicator functioning			
Engine: hydraulic, air, electrical, oil pressure, temperature, and fuel			
Operational Inspection			
Correct counterweight for the load			
Main hoist control functioning			
Auxiliary hoist control functioning			
Anti-two block in place and functioning			
Swing brake			
Lights and horns functional			

Notes:

Date_____

Operator		Signature	
Crane number		Model	

Visual Inspection	Pass	Fail	N/A		Pass	Fail	N/A
Engine fluid level correct (check dip stick or sight glass)				**Gauges and Indicators**			
Hydraulic fluid level correct (check dip stick or sight glass)				Load moment indicator operational			
Hydraulic system exhibits no apparent weeping or leaks				Drum rotation indicator functioning			
Air system exhibits no audible leaks				Boom length indicator functioning			
Tire pressure acceptable and tire not damaged				Boom angle indicator functioning			
Telescoping boom exhibits no damage to structure, wear pads, boom stops, or cylinder				Engine: hydraulic, air, electrical, oil pressure, temperature, and fuel			
Wire rope free of dirt, excess lube, kinks, and wires and spooled correctly				**Operational Inspection**			
Reeving correct				Correct counterweight for the load			
Wedge sockets and wire rope clips not distorted, cracked, or missing				Main hoist control functioning			
Block not damaged				Auxiliary hoist control functioning			
Ball and hook is free to swivel and rotate				Anti-two block in place and functioning			
Guards are in place				Swing brake			
Outrigger float(s) secured with pad pin				Lights and horns functional			
Cab							
Handrails in place and not damaged							
Operator's manual in vehicle							
Load chart legible and visible to operator							
Hand signal chart visible to workers							
Charged fire extinguisher in place							
Cab glass not cracked and wipers are functional							

Notes:

Date_____

Operator
Crane number

Signature	
Model	

Visual Inspection	Pass	Fail	N/A
Engine fluid level correct (check dip stick or sight glass)			
Hydraulic fluid level correct (check dip stick or sight glass)			
Hydraulic system exhibits no apparent weeping or leaks			
Air system exhibits no audible leaks			
Tire pressure acceptable and tire not damaged			
Telescoping boom exhibits no damage to structure, wear pads, boom stops, or cylinder			
Wire rope free of dirt, excess lube, kinks, and wires and spooled correctly			
Reeving correct			
Wedge sockets and wire rope clips not distorted, cracked, or missing			
Block not damaged			
Ball and hook is free to swivel and rotate			
Guards are in place			
Outrigger float(s) secured with pad pin			
Cab			
Handrails in place and not damaged			
Operator's manual in vehicle			
Load chart legible and visible to operator			
Hand signal chart visible to workers			
Charged fire extinguisher in place			
Cab glass not cracked and wipers are functional			

	Pass	Fail	N/A
Gauges and Indicators			
Load moment indicator operational			
Drum rotation indicator functioning			
Boom length indicator functioning			
Boom angle indicator functioning			
Engine: hydraulic, air, electrical, oil pressure, temperature, and fuel			
Operational Inspection			
Correct counterweight for the load			
Main hoist control functioning			
Auxiliary hoist control functioning			
Anti-two block in place and functioning			
Swing brake			
Lights and horns functional			

Notes:

Date_____

Operator		Signature	
Crane number		Model	

Visual Inspection	Pass	Fail	N/A
Engine fluid level correct (check dip stick or sight glass)			
Hydraulic fluid level correct (check dip stick or sight glass)			
Hydraulic system exhibits no apparent weeping or leaks			
Air system exhibits no audible leaks			
Tire pressure acceptable and tire not damaged			
Telescoping boom exhibits no damage to structure, wear pads, boom stops, or cylinder			
Wire rope free of dirt, excess lube, kinks, and wires and spooled correctly			
Reeving correct			
Wedge sockets and wire rope clips not distorted, cracked, or missing			
Block not damaged			
Ball and hook is free to swivel and rotate			
Guards are in place			
Outrigger float(s) secured with pad pin			
Cab			
Handrails in place and not damaged			
Operator's manual in vehicle			
Load chart legible and visible to operator			
Hand signal chart visible to workers			
Charged fire extinguisher in place			
Cab glass not cracked and wipers are functional			

	Pass	Fail	N/A
Gauges and Indicators			
Load moment indicator operational			
Drum rotation indicator functioning			
Boom length indicator functioning			
Boom angle indicator functioning			
Engine: hydraulic, air, electrical, oil pressure, temperature, and fuel			
Operational Inspection			
Correct counterweight for the load			
Main hoist control functioning			
Auxiliary hoist control functioning			
Anti-two block in place and functioning			
Swing brake			
Lights and horns functional			

Notes:

Date_____

Operator
Crane number

Signature	
Model	

Visual Inspection	Pass	Fail	N/A
Engine fluid level correct (check dip stick or sight glass)			
Hydraulic fluid level correct (check dip stick or sight glass)			
Hydraulic system exhibits no apparent weeping or leaks			
Air system exhibits no audible leaks			
Tire pressure acceptable and tire not damaged			
Telescoping boom exhibits no damage to structure, wear pads, boom stops, or cylinder			
Wire rope free of dirt, excess lube, kinks, and wires and spooled correctly			
Reeving correct			
Wedge sockets and wire rope clips not distorted, cracked, or missing			
Block not damaged			
Ball and hook is free to swivel and rotate			
Guards are in place			
Outrigger float(s) secured with pad pin			
Cab			
Handrails in place and not damaged			
Operator's manual in vehicle			
Load chart legible and visible to operator			
Hand signal chart visible to workers			
Charged fire extinguisher in place			
Cab glass not cracked and wipers are functional			

	Pass	Fail	N/A
Gauges and Indicators			
Load moment indicator operational			
Drum rotation indicator functioning			
Boom length indicator functioning			
Boom angle indicator functioning			
Engine: hydraulic, air, electrical, oil pressure, temperature, and fuel			
Operational Inspection			
Correct counterweight for the load			
Main hoist control functioning			
Auxiliary hoist control functioning			
Anti-two block in place and functioning			
Swing brake			
Lights and horns functional			

Notes:

Date_____

Operator		Signature	
Crane number		Model	

Visual Inspection	Pass	Fail	N/A		Pass	Fail	N/A
Engine fluid level correct (check dip stick or sight glass)				**Gauges and Indicators**			
Hydraulic fluid level correct (check dip stick or sight glass)				Load moment indicator operational			
Hydraulic system exhibits no apparent weeping or leaks				Drum rotation indicator functioning			
Air system exhibits no audible leaks				Boom length indicator functioning			
Tire pressure acceptable and tire not damaged				Boom angle indicator functioning			
Telescoping boom exhibits no damage to structure, wear pads, boom stops, or cylinder				Engine: hydraulic, air, electrical, oil pressure, temperature, and fuel			
Wire rope free of dirt, excess lube, kinks, and wires and spooled correctly				**Operational Inspection**			
Reeving correct				Correct counterweight for the load			
Wedge sockets and wire rope clips not distorted, cracked, or missing				Main hoist control functioning			
Block not damaged				Auxiliary hoist control functioning			
Ball and hook is free to swivel and rotate				Anti-two block in place and functioning			
Guards are in place				Swing brake			
Outrigger float(s) secured with pad pin				Lights and horns functional			
Cab							
Handrails in place and not damaged							
Operator's manual in vehicle							
Load chart legible and visible to operator							
Hand signal chart visible to workers							
Charged fire extinguisher in place							
Cab glass not cracked and wipers are functional							

Notes:

Date_____

Operator
Crane number

Signature	
Model	

Visual Inspection	Pass	Fail	N/A
Engine fluid level correct (check dip stick or sight glass)			
Hydraulic fluid level correct (check dip stick or sight glass)			
Hydraulic system exhibits no apparent weeping or leaks			
Air system exhibits no audible leaks			
Tire pressure acceptable and tire not damaged			
Telescoping boom exhibits no damage to structure, wear pads, boom stops, or cylinder			
Wire rope free of dirt, excess lube, kinks, and wires and spooled correctly			
Reeving correct			
Wedge sockets and wire rope clips not distorted, cracked, or missing			
Block not damaged			
Ball and hook is free to swivel and rotate			
Guards are in place			
Outrigger float(s) secured with pad pin			
Cab			
Handrails in place and not damaged			
Operator's manual in vehicle			
Load chart legible and visible to operator			
Hand signal chart visible to workers			
Charged fire extinguisher in place			
Cab glass not cracked and wipers are functional			

	Pass	Fail	N/A
Gauges and Indicators			
Load moment indicator operational			
Drum rotation indicator functioning			
Boom length indicator functioning			
Boom angle indicator functioning			
Engine: hydraulic, air, electrical, oil pressure, temperature, and fuel			
Operational Inspection			
Correct counterweight for the load			
Main hoist control functioning			
Auxiliary hoist control functioning			
Anti-two block in place and functioning			
Swing brake			
Lights and horns functional			

Notes:

Date_____

Operator		Signature	
Crane number		Model	

Visual Inspection	Pass	Fail	N/A
Engine fluid level correct (check dip stick or sight glass)			
Hydraulic fluid level correct (check dip stick or sight glass)			
Hydraulic system exhibits no apparent weeping or leaks			
Air system exhibits no audible leaks			
Tire pressure acceptable and tire not damaged			
Telescoping boom exhibits no damage to structure, wear pads, boom stops, or cylinder			
Wire rope free of dirt, excess lube, kinks, and wires and spooled correctly			
Reeving correct			
Wedge sockets and wire rope clips not distorted, cracked, or missing			
Block not damaged			
Ball and hook is free to swivel and rotate			
Guards are in place			
Outrigger float(s) secured with pad pin			
Cab			
Handrails in place and not damaged			
Operator's manual in vehicle			
Load chart legible and visible to operator			
Hand signal chart visible to workers			
Charged fire extinguisher in place			
Cab glass not cracked and wipers are functional			

	Pass	Fail	N/A
Gauges and Indicators			
Load moment indicator operational			
Drum rotation indicator functioning			
Boom length indicator functioning			
Boom angle indicator functioning			
Engine: hydraulic, air, electrical, oil pressure, temperature, and fuel			
Operational Inspection			
Correct counterweight for the load			
Main hoist control functioning			
Auxiliary hoist control functioning			
Anti-two block in place and functioning			
Swing brake			
Lights and horns functional			

Notes:

Date_____

Operator
Crane number

Signature	
Model	

Visual Inspection	Pass	Fail	N/A
Engine fluid level correct (check dip stick or sight glass)			
Hydraulic fluid level correct (check dip stick or sight glass)			
Hydraulic system exhibits no apparent weeping or leaks			
Air system exhibits no audible leaks			
Tire pressure acceptable and tire not damaged			
Telescoping boom exhibits no damage to structure, wear pads, boom stops, or cylinder			
Wire rope free of dirt, excess lube, kinks, and wires and spooled correctly			
Reeving correct			
Wedge sockets and wire rope clips not distorted, cracked, or missing			
Block not damaged			
Ball and hook is free to swivel and rotate			
Guards are in place			
Outrigger float(s) secured with pad pin			
Cab			
Handrails in place and not damaged			
Operator's manual in vehicle			
Load chart legible and visible to operator			
Hand signal chart visible to workers			
Charged fire extinguisher in place			
Cab glass not cracked and wipers are functional			

	Pass	Fail	N/A
Gauges and Indicators			
Load moment indicator operational			
Drum rotation indicator functioning			
Boom length indicator functioning			
Boom angle indicator functioning			
Engine: hydraulic, air, electrical, oil pressure, temperature, and fuel			
Operational Inspection			
Correct counterweight for the load			
Main hoist control functioning			
Auxiliary hoist control functioning			
Anti-two block in place and functioning			
Swing brake			
Lights and horns functional			

Notes:

Date_____

Operator		Signature	
Crane number		Model	

Visual Inspection	Pass	Fail	N/A		Pass	Fail	N/A
Engine fluid level correct (check dip stick or sight glass)				**Gauges and Indicators**			
Hydraulic fluid level correct (check dip stick or sight glass)				Load moment indicator operational			
Hydraulic system exhibits no apparent weeping or leaks				Drum rotation indicator functioning			
Air system exhibits no audible leaks				Boom length indicator functioning			
Tire pressure acceptable and tire not damaged				Boom angle indicator functioning			
Telescoping boom exhibits no damage to structure, wear pads, boom stops, or cylinder				Engine: hydraulic, air, electrical, oil pressure, temperature, and fuel			
Wire rope free of dirt, excess lube, kinks, and wires and spooled correctly				**Operational Inspection**			
Reeving correct				Correct counterweight for the load			
Wedge sockets and wire rope clips not distorted, cracked, or missing				Main hoist control functioning			
Block not damaged				Auxiliary hoist control functioning			
Ball and hook is free to swivel and rotate				Anti-two block in place and functioning			
Guards are in place				Swing brake			
Outrigger float(s) secured with pad pin				Lights and horns functional			
Cab							
Handrails in place and not damaged							
Operator's manual in vehicle							
Load chart legible and visible to operator							
Hand signal chart visible to workers							
Charged fire extinguisher in place							
Cab glass not cracked and wipers are functional							

Notes:

Date _____

Operator
Crane number

Signature	
Model	

Visual Inspection	Pass	Fail	N/A
Engine fluid level correct (check dip stick or sight glass)			
Hydraulic fluid level correct (check dip stick or sight glass)			
Hydraulic system exhibits no apparent weeping or leaks			
Air system exhibits no audible leaks			
Tire pressure acceptable and tire not damaged			
Telescoping boom exhibits no damage to structure, wear pads, boom stops, or cylinder			
Wire rope free of dirt, excess lube, kinks, and wires and spooled correctly			
Reeving correct			
Wedge sockets and wire rope clips not distorted, cracked, or missing			
Block not damaged			
Ball and hook is free to swivel and rotate			
Guards are in place			
Outrigger float(s) secured with pad pin			
Cab			
Handrails in place and not damaged			
Operator's manual in vehicle			
Load chart legible and visible to operator			
Hand signal chart visible to workers			
Charged fire extinguisher in place			
Cab glass not cracked and wipers are functional			

	Pass	Fail	N/A
Gauges and Indicators			
Load moment indicator operational			
Drum rotation indicator functioning			
Boom length indicator functioning			
Boom angle indicator functioning			
Engine: hydraulic, air, electrical, oil pressure, temperature, and fuel			
Operational Inspection			
Correct counterweight for the load			
Main hoist control functioning			
Auxiliary hoist control functioning			
Anti-two block in place and functioning			
Swing brake			
Lights and horns functional			

Notes:

Date_____

Operator		Signature	
Crane number		Model	

Visual Inspection	Pass	Fail	N/A
Engine fluid level correct (check dip stick or sight glass)			
Hydraulic fluid level correct (check dip stick or sight glass)			
Hydraulic system exhibits no apparent weeping or leaks			
Air system exhibits no audible leaks			
Tire pressure acceptable and tire not damaged			
Telescoping boom exhibits no damage to structure, wear pads, boom stops, or cylinder			
Wire rope free of dirt, excess lube, kinks, and wires and spooled correctly			
Reeving correct			
Wedge sockets and wire rope clips not distorted, cracked, or missing			
Block not damaged			
Ball and hook is free to swivel and rotate			
Guards are in place			
Outrigger float(s) secured with pad pin			
Cab			
Handrails in place and not damaged			
Operator's manual in vehicle			
Load chart legible and visible to operator			
Hand signal chart visible to workers			
Charged fire extinguisher in place			
Cab glass not cracked and wipers are functional			

	Pass	Fail	N/A
Gauges and Indicators			
Load moment indicator operational			
Drum rotation indicator functioning			
Boom length indicator functioning			
Boom angle indicator functioning			
Engine: hydraulic, air, electrical, oil pressure, temperature, and fuel			
Operational Inspection			
Correct counterweight for the load			
Main hoist control functioning			
Auxiliary hoist control functioning			
Anti-two block in place and functioning			
Swing brake			
Lights and horns functional			

Notes:

Date_____

Operator	
Crane number	

Signature	
Model	

Visual Inspection	Pass	Fail	N/A
Engine fluid level correct (check dip stick or sight glass)			
Hydraulic fluid level correct (check dip stick or sight glass)			
Hydraulic system exhibits no apparent weeping or leaks			
Air system exhibits no audible leaks			
Tire pressure acceptable and tire not damaged			
Telescoping boom exhibits no damage to structure, wear pads, boom stops, or cylinder			
Wire rope free of dirt, excess lube, kinks, and wires and spooled correctly			
Reeving correct			
Wedge sockets and wire rope clips not distorted, cracked, or missing			
Block not damaged			
Ball and hook is free to swivel and rotate			
Guards are in place			
Outrigger float(s) secured with pad pin			
Cab			
Handrails in place and not damaged			
Operator's manual in vehicle			
Load chart legible and visible to operator			
Hand signal chart visible to workers			
Charged fire extinguisher in place			
Cab glass not cracked and wipers are functional			

	Pass	Fail	N/A
Gauges and Indicators			
Load moment indicator operational			
Drum rotation indicator functioning			
Boom length indicator functioning			
Boom angle indicator functioning			
Engine: hydraulic, air, electrical, oil pressure, temperature, and fuel			
Operational Inspection			
Correct counterweight for the load			
Main hoist control functioning			
Auxiliary hoist control functioning			
Anti-two block in place and functioning			
Swing brake			
Lights and horns functional			

Notes:

Date_____

Operator		Signature	
Crane number		Model	

Visual Inspection	Pass	Fail	N/A
Engine fluid level correct (check dip stick or sight glass)			
Hydraulic fluid level correct (check dip stick or sight glass)			
Hydraulic system exhibits no apparent weeping or leaks			
Air system exhibits no audible leaks			
Tire pressure acceptable and tire not damaged			
Telescoping boom exhibits no damage to structure, wear pads, boom stops, or cylinder			
Wire rope free of dirt, excess lube, kinks, and wires and spooled correctly			
Reeving correct			
Wedge sockets and wire rope clips not distorted, cracked, or missing			
Block not damaged			
Ball and hook is free to swivel and rotate			
Guards are in place			
Outrigger float(s) secured with pad pin			
Cab			
Handrails in place and not damaged			
Operator's manual in vehicle			
Load chart legible and visible to operator			
Hand signal chart visible to workers			
Charged fire extinguisher in place			
Cab glass not cracked and wipers are functional			

	Pass	Fail	N/A
Gauges and Indicators			
Load moment indicator operational			
Drum rotation indicator functioning			
Boom length indicator functioning			
Boom angle indicator functioning			
Engine: hydraulic, air, electrical, oil pressure, temperature, and fuel			
Operational Inspection			
Correct counterweight for the load			
Main hoist control functioning			
Auxiliary hoist control functioning			
Anti-two block in place and functioning			
Swing brake			
Lights and horns functional			

Notes:

Date_____

Operator
Crane number

Signature	
Model	

Visual Inspection	Pass	Fail	N/A
Engine fluid level correct (check dip stick or sight glass)			
Hydraulic fluid level correct (check dip stick or sight glass)			
Hydraulic system exhibits no apparent weeping or leaks			
Air system exhibits no audible leaks			
Tire pressure acceptable and tire not damaged			
Telescoping boom exhibits no damage to structure, wear pads, boom stops, or cylinder			
Wire rope free of dirt, excess lube, kinks, and wires and spooled correctly			
Reeving correct			
Wedge sockets and wire rope clips not distorted, cracked, or missing			
Block not damaged			
Ball and hook is free to swivel and rotate			
Guards are in place			
Outrigger float(s) secured with pad pin			
Cab			
Handrails in place and not damaged			
Operator's manual in vehicle			
Load chart legible and visible to operator			
Hand signal chart visible to workers			
Charged fire extinguisher in place			
Cab glass not cracked and wipers are functional			

	Pass	Fail	N/A
Gauges and Indicators			
Load moment indicator operational			
Drum rotation indicator functioning			
Boom length indicator functioning			
Boom angle indicator functioning			
Engine: hydraulic, air, electrical, oil pressure, temperature, and fuel			
Operational Inspection			
Correct counterweight for the load			
Main hoist control functioning			
Auxiliary hoist control functioning			
Anti-two block in place and functioning			
Swing brake			
Lights and horns functional			

Notes:

Date_____

Operator		Signature	
Crane number		Model	

Visual Inspection	Pass	Fail	N/A
Engine fluid level correct (check dip stick or sight glass)			
Hydraulic fluid level correct (check dip stick or sight glass)			
Hydraulic system exhibits no apparent weeping or leaks			
Air system exhibits no audible leaks			
Tire pressure acceptable and tire not damaged			
Telescoping boom exhibits no damage to structure, wear pads, boom stops, or cylinder			
Wire rope free of dirt, excess lube, kinks, and wires and spooled correctly			
Reeving correct			
Wedge sockets and wire rope clips not distorted, cracked, or missing			
Block not damaged			
Ball and hook is free to swivel and rotate			
Guards are in place			
Outrigger float(s) secured with pad pin			
Cab			
Handrails in place and not damaged			
Operator's manual in vehicle			
Load chart legible and visible to operator			
Hand signal chart visible to workers			
Charged fire extinguisher in place			
Cab glass not cracked and wipers are functional			

	Pass	Fail	N/A
Gauges and Indicators			
Load moment indicator operational			
Drum rotation indicator functioning			
Boom length indicator functioning			
Boom angle indicator functioning			
Engine: hydraulic, air, electrical, oil pressure, temperature, and fuel			
Operational Inspection			
Correct counterweight for the load			
Main hoist control functioning			
Auxiliary hoist control functioning			
Anti-two block in place and functioning			
Swing brake			
Lights and horns functional			

Notes:

Date_____

Operator
Crane number

Signature	
Model	

Visual Inspection	Pass	Fail	N/A
Engine fluid level correct (check dip stick or sight glass)			
Hydraulic fluid level correct (check dip stick or sight glass)			
Hydraulic system exhibits no apparent weeping or leaks			
Air system exhibits no audible leaks			
Tire pressure acceptable and tire not damaged			
Telescoping boom exhibits no damage to structure, wear pads, boom stops, or cylinder			
Wire rope free of dirt, excess lube, kinks, and wires and spooled correctly			
Reeving correct			
Wedge sockets and wire rope clips not distorted, cracked, or missing			
Block not damaged			
Ball and hook is free to swivel and rotate			
Guards are in place			
Outrigger float(s) secured with pad pin			
Cab			
Handrails in place and not damaged			
Operator's manual in vehicle			
Load chart legible and visible to operator			
Hand signal chart visible to workers			
Charged fire extinguisher in place			
Cab glass not cracked and wipers are functional			

	Pass	Fail	N/A
Gauges and Indicators			
Load moment indicator operational			
Drum rotation indicator functioning			
Boom length indicator functioning			
Boom angle indicator functioning			
Engine: hydraulic, air, electrical, oil pressure, temperature, and fuel			
Operational Inspection			
Correct counterweight for the load			
Main hoist control functioning			
Auxiliary hoist control functioning			
Anti-two block in place and functioning			
Swing brake			
Lights and horns functional			

Notes:

Date_____

Operator		Signature	
Crane number		Model	

Visual Inspection	Pass	Fail	N/A		Pass	Fail	N/A
Engine fluid level correct (check dip stick or sight glass)				**Gauges and Indicators**			
Hydraulic fluid level correct (check dip stick or sight glass)				Load moment indicator operational			
Hydraulic system exhibits no apparent weeping or leaks				Drum rotation indicator functioning			
Air system exhibits no audible leaks				Boom length indicator functioning			
Tire pressure acceptable and tire not damaged				Boom angle indicator functioning			
Telescoping boom exhibits no damage to structure, wear pads, boom stops, or cylinder				Engine: hydraulic, air, electrical, oil pressure, temperature, and fuel			
Wire rope free of dirt, excess lube, kinks, and wires and spooled correctly				**Operational Inspection**			
Reeving correct				Correct counterweight for the load			
Wedge sockets and wire rope clips not distorted, cracked, or missing				Main hoist control functioning			
Block not damaged				Auxiliary hoist control functioning			
Ball and hook is free to swivel and rotate				Anti-two block in place and functioning			
Guards are in place				Swing brake			
Outrigger float(s) secured with pad pin				Lights and horns functional			
Cab							
Handrails in place and not damaged							
Operator's manual in vehicle							
Load chart legible and visible to operator							
Hand signal chart visible to workers							
Charged fire extinguisher in place							
Cab glass not cracked and wipers are functional							

Notes:

Date_____

Operator
Crane number

Signature	
Model	

Visual Inspection	Pass	Fail	N/A
Engine fluid level correct (check dip stick or sight glass)			
Hydraulic fluid level correct (check dip stick or sight glass)			
Hydraulic system exhibits no apparent weeping or leaks			
Air system exhibits no audible leaks			
Tire pressure acceptable and tire not damaged			
Telescoping boom exhibits no damage to structure, wear pads, boom stops, or cylinder			
Wire rope free of dirt, excess lube, kinks, and wires and spooled correctly			
Reeving correct			
Wedge sockets and wire rope clips not distorted, cracked, or missing			
Block not damaged			
Ball and hook is free to swivel and rotate			
Guards are in place			
Outrigger float(s) secured with pad pin			
Cab			
Handrails in place and not damaged			
Operator's manual in vehicle			
Load chart legible and visible to operator			
Hand signal chart visible to workers			
Charged fire extinguisher in place			
Cab glass not cracked and wipers are functional			

	Pass	Fail	N/A
Gauges and Indicators			
Load moment indicator operational			
Drum rotation indicator functioning			
Boom length indicator functioning			
Boom angle indicator functioning			
Engine: hydraulic, air, electrical, oil pressure, temperature, and fuel			
Operational Inspection			
Correct counterweight for the load			
Main hoist control functioning			
Auxiliary hoist control functioning			
Anti-two block in place and functioning			
Swing brake			
Lights and horns functional			

Notes:

Date_____

Operator		Signature	
Crane number		Model	

Visual Inspection	Pass	Fail	N/A		Pass	Fail	N/A
Engine fluid level correct (check dip stick or sight glass)				**Gauges and Indicators**			
Hydraulic fluid level correct (check dip stick or sight glass)				Load moment indicator operational			
Hydraulic system exhibits no apparent weeping or leaks				Drum rotation indicator functioning			
Air system exhibits no audible leaks				Boom length indicator functioning			
Tire pressure acceptable and tire not damaged				Boom angle indicator functioning			
Telescoping boom exhibits no damage to structure, wear pads, boom stops, or cylinder				Engine: hydraulic, air, electrical, oil pressure, temperature, and fuel			
Wire rope free of dirt, excess lube, kinks, and wires and spooled correctly				**Operational Inspection**			
Reeving correct				Correct counterweight for the load			
Wedge sockets and wire rope clips not distorted, cracked, or missing				Main hoist control functioning			
Block not damaged				Auxiliary hoist control functioning			
Ball and hook is free to swivel and rotate				Anti-two block in place and functioning			
Guards are in place				Swing brake			
Outrigger float(s) secured with pad pin				Lights and horns functional			
Cab							
Handrails in place and not damaged							
Operator's manual in vehicle							
Load chart legible and visible to operator							
Hand signal chart visible to workers							
Charged fire extinguisher in place							
Cab glass not cracked and wipers are functional							

Notes:

Date_____

Operator
Crane number

Signature	
Model	

Visual Inspection	Pass	Fail	N/A
Engine fluid level correct (check dip stick or sight glass)			
Hydraulic fluid level correct (check dip stick or sight glass)			
Hydraulic system exhibits no apparent weeping or leaks			
Air system exhibits no audible leaks			
Tire pressure acceptable and tire not damaged			
Telescoping boom exhibits no damage to structure, wear pads, boom stops, or cylinder			
Wire rope free of dirt, excess lube, kinks, and wires and spooled correctly			
Reeving correct			
Wedge sockets and wire rope clips not distorted, cracked, or missing			
Block not damaged			
Ball and hook is free to swivel and rotate			
Guards are in place			
Outrigger float(s) secured with pad pin			
Cab			
Handrails in place and not damaged			
Operator's manual in vehicle			
Load chart legible and visible to operator			
Hand signal chart visible to workers			
Charged fire extinguisher in place			
Cab glass not cracked and wipers are functional			

	Pass	Fail	N/A
Gauges and Indicators			
Load moment indicator operational			
Drum rotation indicator functioning			
Boom length indicator functioning			
Boom angle indicator functioning			
Engine: hydraulic, air, electrical, oil pressure, temperature, and fuel			
Operational Inspection			
Correct counterweight for the load			
Main hoist control functioning			
Auxiliary hoist control functioning			
Anti-two block in place and functioning			
Swing brake			
Lights and horns functional			

Notes:

Date_____

Operator		Signature	
Crane number		Model	

Visual Inspection	Pass	Fail	N/A
Engine fluid level correct (check dip stick or sight glass)			
Hydraulic fluid level correct (check dip stick or sight glass)			
Hydraulic system exhibits no apparent weeping or leaks			
Air system exhibits no audible leaks			
Tire pressure acceptable and tire not damaged			
Telescoping boom exhibits no damage to structure, wear pads, boom stops, or cylinder			
Wire rope free of dirt, excess lube, kinks, and wires and spooled correctly			
Reeving correct			
Wedge sockets and wire rope clips not distorted, cracked, or missing			
Block not damaged			
Ball and hook is free to swivel and rotate			
Guards are in place			
Outrigger float(s) secured with pad pin			
Cab			
Handrails in place and not damaged			
Operator's manual in vehicle			
Load chart legible and visible to operator			
Hand signal chart visible to workers			
Charged fire extinguisher in place			
Cab glass not cracked and wipers are functional			

	Pass	Fail	N/A
Gauges and Indicators			
Load moment indicator operational			
Drum rotation indicator functioning			
Boom length indicator functioning			
Boom angle indicator functioning			
Engine: hydraulic, air, electrical, oil pressure, temperature, and fuel			
Operational Inspection			
Correct counterweight for the load			
Main hoist control functioning			
Auxiliary hoist control functioning			
Anti-two block in place and functioning			
Swing brake			
Lights and horns functional			

Notes:

Date_____

Operator
Crane number

Signature	
Model	

Visual Inspection	Pass	Fail	N/A
Engine fluid level correct (check dip stick or sight glass)			
Hydraulic fluid level correct (check dip stick or sight glass)			
Hydraulic system exhibits no apparent weeping or leaks			
Air system exhibits no audible leaks			
Tire pressure acceptable and tire not damaged			
Telescoping boom exhibits no damage to structure, wear pads, boom stops, or cylinder			
Wire rope free of dirt, excess lube, kinks, and wires and spooled correctly			
Reeving correct			
Wedge sockets and wire rope clips not distorted, cracked, or missing			
Block not damaged			
Ball and hook is free to swivel and rotate			
Guards are in place			
Outrigger float(s) secured with pad pin			
Cab			
Handrails in place and not damaged			
Operator's manual in vehicle			
Load chart legible and visible to operator			
Hand signal chart visible to workers			
Charged fire extinguisher in place			
Cab glass not cracked and wipers are functional			

	Pass	Fail	N/A
Gauges and Indicators			
Load moment indicator operational			
Drum rotation indicator functioning			
Boom length indicator functioning			
Boom angle indicator functioning			
Engine: hydraulic, air, electrical, oil pressure, temperature, and fuel			
Operational Inspection			
Correct counterweight for the load			
Main hoist control functioning			
Auxiliary hoist control functioning			
Anti-two block in place and functioning			
Swing brake			
Lights and horns functional			

Notes:

Date_____

Operator		Signature	
Crane number		Model	

Visual Inspection	Pass	Fail	N/A
Engine fluid level correct (check dip stick or sight glass)			
Hydraulic fluid level correct (check dip stick or sight glass)			
Hydraulic system exhibits no apparent weeping or leaks			
Air system exhibits no audible leaks			
Tire pressure acceptable and tire not damaged			
Telescoping boom exhibits no damage to structure, wear pads, boom stops, or cylinder			
Wire rope free of dirt, excess lube, kinks, and wires and spooled correctly			
Reeving correct			
Wedge sockets and wire rope clips not distorted, cracked, or missing			
Block not damaged			
Ball and hook is free to swivel and rotate			
Guards are in place			
Outrigger float(s) secured with pad pin			
Cab			
Handrails in place and not damaged			
Operator's manual in vehicle			
Load chart legible and visible to operator			
Hand signal chart visible to workers			
Charged fire extinguisher in place			
Cab glass not cracked and wipers are functional			

	Pass	Fail	N/A
Gauges and Indicators			
Load moment indicator operational			
Drum rotation indicator functioning			
Boom length indicator functioning			
Boom angle indicator functioning			
Engine: hydraulic, air, electrical, oil pressure, temperature, and fuel			
Operational Inspection			
Correct counterweight for the load			
Main hoist control functioning			
Auxiliary hoist control functioning			
Anti-two block in place and functioning			
Swing brake			
Lights and horns functional			

Notes:

Date_____

Operator
Crane number

Signature	
Model	

Visual Inspection	Pass	Fail	N/A
Engine fluid level correct (check dip stick or sight glass)			
Hydraulic fluid level correct (check dip stick or sight glass)			
Hydraulic system exhibits no apparent weeping or leaks			
Air system exhibits no audible leaks			
Tire pressure acceptable and tire not damaged			
Telescoping boom exhibits no damage to structure, wear pads, boom stops, or cylinder			
Wire rope free of dirt, excess lube, kinks, and wires and spooled correctly			
Reeving correct			
Wedge sockets and wire rope clips not distorted, cracked, or missing			
Block not damaged			
Ball and hook is free to swivel and rotate			
Guards are in place			
Outrigger float(s) secured with pad pin			
Cab			
Handrails in place and not damaged			
Operator's manual in vehicle			
Load chart legible and visible to operator			
Hand signal chart visible to workers			
Charged fire extinguisher in place			
Cab glass not cracked and wipers are functional			

	Pass	Fail	N/A
Gauges and Indicators			
Load moment indicator operational			
Drum rotation indicator functioning			
Boom length indicator functioning			
Boom angle indicator functioning			
Engine: hydraulic, air, electrical, oil pressure, temperature, and fuel			
Operational Inspection			
Correct counterweight for the load			
Main hoist control functioning			
Auxiliary hoist control functioning			
Anti-two block in place and functioning			
Swing brake			
Lights and horns functional			

Notes:

Date_____

Operator		Signature	
Crane number		Model	

Visual Inspection	Pass	Fail	N/A
Engine fluid level correct (check dip stick or sight glass)			
Hydraulic fluid level correct (check dip stick or sight glass)			
Hydraulic system exhibits no apparent weeping or leaks			
Air system exhibits no audible leaks			
Tire pressure acceptable and tire not damaged			
Telescoping boom exhibits no damage to structure, wear pads, boom stops, or cylinder			
Wire rope free of dirt, excess lube, kinks, and wires and spooled correctly			
Reeving correct			
Wedge sockets and wire rope clips not distorted, cracked, or missing			
Block not damaged			
Ball and hook is free to swivel and rotate			
Guards are in place			
Outrigger float(s) secured with pad pin			
Cab			
Handrails in place and not damaged			
Operator's manual in vehicle			
Load chart legible and visible to operator			
Hand signal chart visible to workers			
Charged fire extinguisher in place			
Cab glass not cracked and wipers are functional			

	Pass	Fail	N/A
Gauges and Indicators			
Load moment indicator operational			
Drum rotation indicator functioning			
Boom length indicator functioning			
Boom angle indicator functioning			
Engine: hydraulic, air, electrical, oil pressure, temperature, and fuel			
Operational Inspection			
Correct counterweight for the load			
Main hoist control functioning			
Auxiliary hoist control functioning			
Anti-two block in place and functioning			
Swing brake			
Lights and horns functional			

Notes:

Date_____

Operator
Crane number

Signature	
Model	

Visual Inspection	Pass	Fail	N/A
Engine fluid level correct (check dip stick or sight glass)			
Hydraulic fluid level correct (check dip stick or sight glass)			
Hydraulic system exhibits no apparent weeping or leaks			
Air system exhibits no audible leaks			
Tire pressure acceptable and tire not damaged			
Telescoping boom exhibits no damage to structure, wear pads, boom stops, or cylinder			
Wire rope free of dirt, excess lube, kinks, and wires and spooled correctly			
Reeving correct			
Wedge sockets and wire rope clips not distorted, cracked, or missing			
Block not damaged			
Ball and hook is free to swivel and rotate			
Guards are in place			
Outrigger float(s) secured with pad pin			
Cab			
Handrails in place and not damaged			
Operator's manual in vehicle			
Load chart legible and visible to operator			
Hand signal chart visible to workers			
Charged fire extinguisher in place			
Cab glass not cracked and wipers are functional			

	Pass	Fail	N/A
Gauges and Indicators			
Load moment indicator operational			
Drum rotation indicator functioning			
Boom length indicator functioning			
Boom angle indicator functioning			
Engine: hydraulic, air, electrical, oil pressure, temperature, and fuel			
Operational Inspection			
Correct counterweight for the load			
Main hoist control functioning			
Auxiliary hoist control functioning			
Anti-two block in place and functioning			
Swing brake			
Lights and horns functional			

Notes:

Date_____

Operator		Signature	
Crane number		Model	

Visual Inspection	Pass	Fail	N/A
Engine fluid level correct (check dip stick or sight glass)			
Hydraulic fluid level correct (check dip stick or sight glass)			
Hydraulic system exhibits no apparent weeping or leaks			
Air system exhibits no audible leaks			
Tire pressure acceptable and tire not damaged			
Telescoping boom exhibits no damage to structure, wear pads, boom stops, or cylinder			
Wire rope free of dirt, excess lube, kinks, and wires and spooled correctly			
Reeving correct			
Wedge sockets and wire rope clips not distorted, cracked, or missing			
Block not damaged			
Ball and hook is free to swivel and rotate			
Guards are in place			
Outrigger float(s) secured with pad pin			
Cab			
Handrails in place and not damaged			
Operator's manual in vehicle			
Load chart legible and visible to operator			
Hand signal chart visible to workers			
Charged fire extinguisher in place			
Cab glass not cracked and wipers are functional			

	Pass	Fail	N/A
Gauges and Indicators			
Load moment indicator operational			
Drum rotation indicator functioning			
Boom length indicator functioning			
Boom angle indicator functioning			
Engine: hydraulic, air, electrical, oil pressure, temperature, and fuel			
Operational Inspection			
Correct counterweight for the load			
Main hoist control functioning			
Auxiliary hoist control functioning			
Anti-two block in place and functioning			
Swing brake			
Lights and horns functional			

Notes:

Date _____

Operator		Signature	
Crane number		Model	

Visual Inspection	Pass	Fail	N/A
Engine fluid level correct (check dip stick or sight glass)			
Hydraulic fluid level correct (check dip stick or sight glass)			
Hydraulic system exhibits no apparent weeping or leaks			
Air system exhibits no audible leaks			
Tire pressure acceptable and tire not damaged			
Telescoping boom exhibits no damage to structure, wear pads, boom stops, or cylinder			
Wire rope free of dirt, excess lube, kinks, and wires and spooled correctly			
Reeving correct			
Wedge sockets and wire rope clips not distorted, cracked, or missing			
Block not damaged			
Ball and hook is free to swivel and rotate			
Guards are in place			
Outrigger float(s) secured with pad pin			
Cab			
Handrails in place and not damaged			
Operator's manual in vehicle			
Load chart legible and visible to operator			
Hand signal chart visible to workers			
Charged fire extinguisher in place			
Cab glass not cracked and wipers are functional			

	Pass	Fail	N/A
Gauges and Indicators			
Load moment indicator operational			
Drum rotation indicator functioning			
Boom length indicator functioning			
Boom angle indicator functioning			
Engine: hydraulic, air, electrical, oil pressure, temperature, and fuel			
Operational Inspection			
Correct counterweight for the load			
Main hoist control functioning			
Auxiliary hoist control functioning			
Anti-two block in place and functioning			
Swing brake			
Lights and horns functional			

Notes:

Date_____

Operator		Signature	
Crane number		Model	

Visual Inspection	Pass	Fail	N/A		Pass	Fail	N/A
Engine fluid level correct (check dip stick or sight glass)				**Gauges and Indicators**			
Hydraulic fluid level correct (check dip stick or sight glass)				Load moment indicator operational			
Hydraulic system exhibits no apparent weeping or leaks				Drum rotation indicator functioning			
Air system exhibits no audible leaks				Boom length indicator functioning			
Tire pressure acceptable and tire not damaged				Boom angle indicator functioning			
Telescoping boom exhibits no damage to structure, wear pads, boom stops, or cylinder				Engine: hydraulic, air, electrical, oil pressure, temperature, and fuel			
Wire rope free of dirt, excess lube, kinks, and wires and spooled correctly				**Operational Inspection**			
Reeving correct				Correct counterweight for the load			
Wedge sockets and wire rope clips not distorted, cracked, or missing				Main hoist control functioning			
Block not damaged				Auxiliary hoist control functioning			
Ball and hook is free to swivel and rotate				Anti-two block in place and functioning			
Guards are in place				Swing brake			
Outrigger float(s) secured with pad pin				Lights and horns functional			
Cab							
Handrails in place and not damaged							
Operator's manual in vehicle							
Load chart legible and visible to operator							
Hand signal chart visible to workers							
Charged fire extinguisher in place							
Cab glass not cracked and wipers are functional							

Notes:

Date_____

Operator		Signature	
Crane number		Model	

Visual Inspection	Pass	Fail	N/A
Engine fluid level correct (check dip stick or sight glass)			
Hydraulic fluid level correct (check dip stick or sight glass)			
Hydraulic system exhibits no apparent weeping or leaks			
Air system exhibits no audible leaks			
Tire pressure acceptable and tire not damaged			
Telescoping boom exhibits no damage to structure, wear pads, boom stops, or cylinder			
Wire rope free of dirt, excess lube, kinks, and wires and spooled correctly			
Reeving correct			
Wedge sockets and wire rope clips not distorted, cracked, or missing			
Block not damaged			
Ball and hook is free to swivel and rotate			
Guards are in place			
Outrigger float(s) secured with pad pin			
Cab			
Handrails in place and not damaged			
Operator's manual in vehicle			
Load chart legible and visible to operator			
Hand signal chart visible to workers			
Charged fire extinguisher in place			
Cab glass not cracked and wipers are functional			

	Pass	Fail	N/A
Gauges and Indicators			
Load moment indicator operational			
Drum rotation indicator functioning			
Boom length indicator functioning			
Boom angle indicator functioning			
Engine: hydraulic, air, electrical, oil pressure, temperature, and fuel			
Operational Inspection			
Correct counterweight for the load			
Main hoist control functioning			
Auxiliary hoist control functioning			
Anti-two block in place and functioning			
Swing brake			
Lights and horns functional			

Notes:

Date_____

Operator		Signature	
Crane number		Model	

Visual Inspection	Pass	Fail	N/A		Pass	Fail	N/A
Engine fluid level correct (check dip stick or sight glass)				**Gauges and Indicators**			
Hydraulic fluid level correct (check dip stick or sight glass)				Load moment indicator operational			
Hydraulic system exhibits no apparent weeping or leaks				Drum rotation indicator functioning			
Air system exhibits no audible leaks				Boom length indicator functioning			
Tire pressure acceptable and tire not damaged				Boom angle indicator functioning			
Telescoping boom exhibits no damage to structure, wear pads, boom stops, or cylinder				Engine: hydraulic, air, electrical, oil pressure, temperature, and fuel			
Wire rope free of dirt, excess lube, kinks, and wires and spooled correctly				**Operational Inspection**			
Reeving correct				Correct counterweight for the load			
Wedge sockets and wire rope clips not distorted, cracked, or missing				Main hoist control functioning			
Block not damaged				Auxiliary hoist control functioning			
Ball and hook is free to swivel and rotate				Anti-two block in place and functioning			
Guards are in place				Swing brake			
Outrigger float(s) secured with pad pin				Lights and horns functional			
Cab							
Handrails in place and not damaged							
Operator's manual in vehicle							
Load chart legible and visible to operator							
Hand signal chart visible to workers							
Charged fire extinguisher in place							
Cab glass not cracked and wipers are functional							

Notes:

Date_____

Operator
Crane number

Signature	
Model	

Visual Inspection	Pass	Fail	N/A
Engine fluid level correct (check dip stick or sight glass)			
Hydraulic fluid level correct (check dip stick or sight glass)			
Hydraulic system exhibits no apparent weeping or leaks			
Air system exhibits no audible leaks			
Tire pressure acceptable and tire not damaged			
Telescoping boom exhibits no damage to structure, wear pads, boom stops, or cylinder			
Wire rope free of dirt, excess lube, kinks, and wires and spooled correctly			
Reeving correct			
Wedge sockets and wire rope clips not distorted, cracked, or missing			
Block not damaged			
Ball and hook is free to swivel and rotate			
Guards are in place			
Outrigger float(s) secured with pad pin			
Cab			
Handrails in place and not damaged			
Operator's manual in vehicle			
Load chart legible and visible to operator			
Hand signal chart visible to workers			
Charged fire extinguisher in place			
Cab glass not cracked and wipers are functional			

Gauges and Indicators	Pass	Fail	N/A
Load moment indicator operational			
Drum rotation indicator functioning			
Boom length indicator functioning			
Boom angle indicator functioning			
Engine: hydraulic, air, electrical, oil pressure, temperature, and fuel			
Operational Inspection			
Correct counterweight for the load			
Main hoist control functioning			
Auxiliary hoist control functioning			
Anti-two block in place and functioning			
Swing brake			
Lights and horns functional			

Notes:

Date_____

Operator		Signature	
Crane number		Model	

Visual Inspection	Pass	Fail	N/A		Pass	Fail	N/A
Engine fluid level correct (check dip stick or sight glass)				**Gauges and Indicators**			
Hydraulic fluid level correct (check dip stick or sight glass)				Load moment indicator operational			
Hydraulic system exhibits no apparent weeping or leaks				Drum rotation indicator functioning			
Air system exhibits no audible leaks				Boom length indicator functioning			
Tire pressure acceptable and tire not damaged				Boom angle indicator functioning			
Telescoping boom exhibits no damage to structure, wear pads, boom stops, or cylinder				Engine: hydraulic, air, electrical, oil pressure, temperature, and fuel			
Wire rope free of dirt, excess lube, kinks, and wires and spooled correctly				**Operational Inspection**			
Reeving correct				Correct counterweight for the load			
Wedge sockets and wire rope clips not distorted, cracked, or missing				Main hoist control functioning			
Block not damaged				Auxiliary hoist control functioning			
Ball and hook is free to swivel and rotate				Anti-two block in place and functioning			
Guards are in place				Swing brake			
Outrigger float(s) secured with pad pin				Lights and horns functional			
Cab							
Handrails in place and not damaged							
Operator's manual in vehicle							
Load chart legible and visible to operator							
Hand signal chart visible to workers							
Charged fire extinguisher in place							
Cab glass not cracked and wipers are functional							

Notes:

Date_____

Operator
Crane number

Signature	
Model	

Visual Inspection	Pass	Fail	N/A
Engine fluid level correct (check dip stick or sight glass)			
Hydraulic fluid level correct (check dip stick or sight glass)			
Hydraulic system exhibits no apparent weeping or leaks			
Air system exhibits no audible leaks			
Tire pressure acceptable and tire not damaged			
Telescoping boom exhibits no damage to structure, wear pads, boom stops, or cylinder			
Wire rope free of dirt, excess lube, kinks, and wires and spooled correctly			
Reeving correct			
Wedge sockets and wire rope clips not distorted, cracked, or missing			
Block not damaged			
Ball and hook is free to swivel and rotate			
Guards are in place			
Outrigger float(s) secured with pad pin			
Cab			
Handrails in place and not damaged			
Operator's manual in vehicle			
Load chart legible and visible to operator			
Hand signal chart visible to workers			
Charged fire extinguisher in place			
Cab glass not cracked and wipers are functional			

	Pass	Fail	N/A
Gauges and Indicators			
Load moment indicator operational			
Drum rotation indicator functioning			
Boom length indicator functioning			
Boom angle indicator functioning			
Engine: hydraulic, air, electrical, oil pressure, temperature, and fuel			
Operational Inspection			
Correct counterweight for the load			
Main hoist control functioning			
Auxiliary hoist control functioning			
Anti-two block in place and functioning			
Swing brake			
Lights and horns functional			

Notes:

Date_____

Operator	
Crane number	

Signature	
Model	

Visual Inspection	Pass	Fail	N/A
Engine fluid level correct (check dip stick or sight glass)			
Hydraulic fluid level correct (check dip stick or sight glass)			
Hydraulic system exhibits no apparent weeping or leaks			
Air system exhibits no audible leaks			
Tire pressure acceptable and tire not damaged			
Telescoping boom exhibits no damage to structure, wear pads, boom stops, or cylinder			
Wire rope free of dirt, excess lube, kinks, and wires and spooled correctly			
Reeving correct			
Wedge sockets and wire rope clips not distorted, cracked, or missing			
Block not damaged			
Ball and hook is free to swivel and rotate			
Guards are in place			
Outrigger float(s) secured with pad pin			
Cab			
Handrails in place and not damaged			
Operator's manual in vehicle			
Load chart legible and visible to operator			
Hand signal chart visible to workers			
Charged fire extinguisher in place			
Cab glass not cracked and wipers are functional			

	Pass	Fail	N/A
Gauges and Indicators			
Load moment indicator operational			
Drum rotation indicator functioning			
Boom length indicator functioning			
Boom angle indicator functioning			
Engine: hydraulic, air, electrical, oil pressure, temperature, and fuel			
Operational Inspection			
Correct counterweight for the load			
Main hoist control functioning			
Auxiliary hoist control functioning			
Anti-two block in place and functioning			
Swing brake			
Lights and horns functional			

Notes:

Date_____

Operator
Crane number

Signature	
Model	

Visual Inspection	Pass	Fail	N/A
Engine fluid level correct (check dip stick or sight glass)			
Hydraulic fluid level correct (check dip stick or sight glass)			
Hydraulic system exhibits no apparent weeping or leaks			
Air system exhibits no audible leaks			
Tire pressure acceptable and tire not damaged			
Telescoping boom exhibits no damage to structure, wear pads, boom stops, or cylinder			
Wire rope free of dirt, excess lube, kinks, and wires and spooled correctly			
Reeving correct			
Wedge sockets and wire rope clips not distorted, cracked, or missing			
Block not damaged			
Ball and hook is free to swivel and rotate			
Guards are in place			
Outrigger float(s) secured with pad pin			
Cab			
Handrails in place and not damaged			
Operator's manual in vehicle			
Load chart legible and visible to operator			
Hand signal chart visible to workers			
Charged fire extinguisher in place			
Cab glass not cracked and wipers are functional			

	Pass	Fail	N/A
Gauges and Indicators			
Load moment indicator operational			
Drum rotation indicator functioning			
Boom length indicator functioning			
Boom angle indicator functioning			
Engine: hydraulic, air, electrical, oil pressure, temperature, and fuel			
Operational Inspection			
Correct counterweight for the load			
Main hoist control functioning			
Auxiliary hoist control functioning			
Anti-two block in place and functioning			
Swing brake			
Lights and horns functional			

Notes:

Date_____

Operator	
Crane number	

Signature		
Model		

Visual Inspection	Pass	Fail	N/A
Engine fluid level correct (check dip stick or sight glass)			
Hydraulic fluid level correct (check dip stick or sight glass)			
Hydraulic system exhibits no apparent weeping or leaks			
Air system exhibits no audible leaks			
Tire pressure acceptable and tire not damaged			
Telescoping boom exhibits no damage to structure, wear pads, boom stops, or cylinder			
Wire rope free of dirt, excess lube, kinks, and wires and spooled correctly			
Reeving correct			
Wedge sockets and wire rope clips not distorted, cracked, or missing			
Block not damaged			
Ball and hook is free to swivel and rotate			
Guards are in place			
Outrigger float(s) secured with pad pin			
Cab			
Handrails in place and not damaged			
Operator's manual in vehicle			
Load chart legible and visible to operator			
Hand signal chart visible to workers			
Charged fire extinguisher in place			
Cab glass not cracked and wipers are functional			

	Pass	Fail	N/A
Gauges and Indicators			
Load moment indicator operational			
Drum rotation indicator functioning			
Boom length indicator functioning			
Boom angle indicator functioning			
Engine: hydraulic, air, electrical, oil pressure, temperature, and fuel			
Operational Inspection			
Correct counterweight for the load			
Main hoist control functioning			
Auxiliary hoist control functioning			
Anti-two block in place and functioning			
Swing brake			
Lights and horns functional			

Notes:

Date_____

Operator
Crane number

Signature	
Model	

Visual Inspection	Pass	Fail	N/A
Engine fluid level correct (check dip stick or sight glass)			
Hydraulic fluid level correct (check dip stick or sight glass)			
Hydraulic system exhibits no apparent weeping or leaks			
Air system exhibits no audible leaks			
Tire pressure acceptable and tire not damaged			
Telescoping boom exhibits no damage to structure, wear pads, boom stops, or cylinder			
Wire rope free of dirt, excess lube, kinks, and wires and spooled correctly			
Reeving correct			
Wedge sockets and wire rope clips not distorted, cracked, or missing			
Block not damaged			
Ball and hook is free to swivel and rotate			
Guards are in place			
Outrigger float(s) secured with pad pin			
Cab			
Handrails in place and not damaged			
Operator's manual in vehicle			
Load chart legible and visible to operator			
Hand signal chart visible to workers			
Charged fire extinguisher in place			
Cab glass not cracked and wipers are functional			

	Pass	Fail	N/A
Gauges and Indicators			
Load moment indicator operational			
Drum rotation indicator functioning			
Boom length indicator functioning			
Boom angle indicator functioning			
Engine: hydraulic, air, electrical, oil pressure, temperature, and fuel			
Operational Inspection			
Correct counterweight for the load			
Main hoist control functioning			
Auxiliary hoist control functioning			
Anti-two block in place and functioning			
Swing brake			
Lights and horns functional			

Notes:

Date_____

Operator		Signature	
Crane number		Model	

Visual Inspection	Pass	Fail	N/A		Pass	Fail	N/A
Engine fluid level correct (check dip stick or sight glass)				**Gauges and Indicators**			
Hydraulic fluid level correct (check dip stick or sight glass)				Load moment indicator operational			
Hydraulic system exhibits no apparent weeping or leaks				Drum rotation indicator functioning			
Air system exhibits no audible leaks				Boom length indicator functioning			
Tire pressure acceptable and tire not damaged				Boom angle indicator functioning			
Telescoping boom exhibits no damage to structure, wear pads, boom stops, or cylinder				Engine: hydraulic, air, electrical, oil pressure, temperature, and fuel			
Wire rope free of dirt, excess lube, kinks, and wires and spooled correctly				**Operational Inspection**			
Reeving correct				Correct counterweight for the load			
Wedge sockets and wire rope clips not distorted, cracked, or missing				Main hoist control functioning			
Block not damaged				Auxiliary hoist control functioning			
Ball and hook is free to swivel and rotate				Anti-two block in place and functioning			
Guards are in place				Swing brake			
Outrigger float(s) secured with pad pin				Lights and horns functional			
Cab							
Handrails in place and not damaged							
Operator's manual in vehicle							
Load chart legible and visible to operator							
Hand signal chart visible to workers							
Charged fire extinguisher in place							
Cab glass not cracked and wipers are functional							

Notes:

Date_____

Operator
Crane number

Signature	
Model	

Visual Inspection	Pass	Fail	N/A
Engine fluid level correct (check dip stick or sight glass)			
Hydraulic fluid level correct (check dip stick or sight glass)			
Hydraulic system exhibits no apparent weeping or leaks			
Air system exhibits no audible leaks			
Tire pressure acceptable and tire not damaged			
Telescoping boom exhibits no damage to structure, wear pads, boom stops, or cylinder			
Wire rope free of dirt, excess lube, kinks, and wires and spooled correctly			
Reeving correct			
Wedge sockets and wire rope clips not distorted, cracked, or missing			
Block not damaged			
Ball and hook is free to swivel and rotate			
Guards are in place			
Outrigger float(s) secured with pad pin			
Cab			
Handrails in place and not damaged			
Operator's manual in vehicle			
Load chart legible and visible to operator			
Hand signal chart visible to workers			
Charged fire extinguisher in place			
Cab glass not cracked and wipers are functional			

	Pass	Fail	N/A
Gauges and Indicators			
Load moment indicator operational			
Drum rotation indicator functioning			
Boom length indicator functioning			
Boom angle indicator functioning			
Engine: hydraulic, air, electrical, oil pressure, temperature, and fuel			
Operational Inspection			
Correct counterweight for the load			
Main hoist control functioning			
Auxiliary hoist control functioning			
Anti-two block in place and functioning			
Swing brake			
Lights and horns functional			

Notes:

Date_____

Operator		Signature	
Crane number		Model	

Visual Inspection	Pass	Fail	N/A
Engine fluid level correct (check dip stick or sight glass)			
Hydraulic fluid level correct (check dip stick or sight glass)			
Hydraulic system exhibits no apparent weeping or leaks			
Air system exhibits no audible leaks			
Tire pressure acceptable and tire not damaged			
Telescoping boom exhibits no damage to structure, wear pads, boom stops, or cylinder			
Wire rope free of dirt, excess lube, kinks, and wires and spooled correctly			
Reeving correct			
Wedge sockets and wire rope clips not distorted, cracked, or missing			
Block not damaged			
Ball and hook is free to swivel and rotate			
Guards are in place			
Outrigger float(s) secured with pad pin			
Cab			
Handrails in place and not damaged			
Operator's manual in vehicle			
Load chart legible and visible to operator			
Hand signal chart visible to workers			
Charged fire extinguisher in place			
Cab glass not cracked and wipers are functional			

	Pass	Fail	N/A
Gauges and Indicators			
Load moment indicator operational			
Drum rotation indicator functioning			
Boom length indicator functioning			
Boom angle indicator functioning			
Engine: hydraulic, air, electrical, oil pressure, temperature, and fuel			
Operational Inspection			
Correct counterweight for the load			
Main hoist control functioning			
Auxiliary hoist control functioning			
Anti-two block in place and functioning			
Swing brake			
Lights and horns functional			

Notes:

Date_____

Operator
Crane number

Signature	
Model	

Visual Inspection	Pass	Fail	N/A
Engine fluid level correct (check dip stick or sight glass)			
Hydraulic fluid level correct (check dip stick or sight glass)			
Hydraulic system exhibits no apparent weeping or leaks			
Air system exhibits no audible leaks			
Tire pressure acceptable and tire not damaged			
Telescoping boom exhibits no damage to structure, wear pads, boom stops, or cylinder			
Wire rope free of dirt, excess lube, kinks, and wires and spooled correctly			
Reeving correct			
Wedge sockets and wire rope clips not distorted, cracked, or missing			
Block not damaged			
Ball and hook is free to swivel and rotate			
Guards are in place			
Outrigger float(s) secured with pad pin			
Cab			
Handrails in place and not damaged			
Operator's manual in vehicle			
Load chart legible and visible to operator			
Hand signal chart visible to workers			
Charged fire extinguisher in place			
Cab glass not cracked and wipers are functional			

	Pass	Fail	N/A
Gauges and Indicators			
Load moment indicator operational			
Drum rotation indicator functioning			
Boom length indicator functioning			
Boom angle indicator functioning			
Engine: hydraulic, air, electrical, oil pressure, temperature, and fuel			
Operational Inspection			
Correct counterweight for the load			
Main hoist control functioning			
Auxiliary hoist control functioning			
Anti-two block in place and functioning			
Swing brake			
Lights and horns functional			

Notes:

Date_____

Operator		Signature	
Crane number		Model	

Visual Inspection	Pass	Fail	N/A
Engine fluid level correct (check dip stick or sight glass)			
Hydraulic fluid level correct (check dip stick or sight glass)			
Hydraulic system exhibits no apparent weeping or leaks			
Air system exhibits no audible leaks			
Tire pressure acceptable and tire not damaged			
Telescoping boom exhibits no damage to structure, wear pads, boom stops, or cylinder			
Wire rope free of dirt, excess lube, kinks, and wires and spooled correctly			
Reeving correct			
Wedge sockets and wire rope clips not distorted, cracked, or missing			
Block not damaged			
Ball and hook is free to swivel and rotate			
Guards are in place			
Outrigger float(s) secured with pad pin			
Cab			
Handrails in place and not damaged			
Operator's manual in vehicle			
Load chart legible and visible to operator			
Hand signal chart visible to workers			
Charged fire extinguisher in place			
Cab glass not cracked and wipers are functional			

	Pass	Fail	N/A
Gauges and Indicators			
Load moment indicator operational			
Drum rotation indicator functioning			
Boom length indicator functioning			
Boom angle indicator functioning			
Engine: hydraulic, air, electrical, oil pressure, temperature, and fuel			
Operational Inspection			
Correct counterweight for the load			
Main hoist control functioning			
Auxiliary hoist control functioning			
Anti-two block in place and functioning			
Swing brake			
Lights and horns functional			

Notes:

Date_____

Operator		Signature	
Crane number		Model	

Visual Inspection	Pass	Fail	N/A
Engine fluid level correct (check dip stick or sight glass)			
Hydraulic fluid level correct (check dip stick or sight glass)			
Hydraulic system exhibits no apparent weeping or leaks			
Air system exhibits no audible leaks			
Tire pressure acceptable and tire not damaged			
Telescoping boom exhibits no damage to structure, wear pads, boom stops, or cylinder			
Wire rope free of dirt, excess lube, kinks, and wires and spooled correctly			
Reeving correct			
Wedge sockets and wire rope clips not distorted, cracked, or missing			
Block not damaged			
Ball and hook is free to swivel and rotate			
Guards are in place			
Outrigger float(s) secured with pad pin			
Cab			
Handrails in place and not damaged			
Operator's manual in vehicle			
Load chart legible and visible to operator			
Hand signal chart visible to workers			
Charged fire extinguisher in place			
Cab glass not cracked and wipers are functional			

	Pass	Fail	N/A
Gauges and Indicators			
Load moment indicator operational			
Drum rotation indicator functioning			
Boom length indicator functioning			
Boom angle indicator functioning			
Engine: hydraulic, air, electrical, oil pressure, temperature, and fuel			
Operational Inspection			
Correct counterweight for the load			
Main hoist control functioning			
Auxiliary hoist control functioning			
Anti-two block in place and functioning			
Swing brake			
Lights and horns functional			

Notes:

Date_____

Operator		Signature	
Crane number		Model	

Visual Inspection	Pass	Fail	N/A
Engine fluid level correct (check dip stick or sight glass)			
Hydraulic fluid level correct (check dip stick or sight glass)			
Hydraulic system exhibits no apparent weeping or leaks			
Air system exhibits no audible leaks			
Tire pressure acceptable and tire not damaged			
Telescoping boom exhibits no damage to structure, wear pads, boom stops, or cylinder			
Wire rope free of dirt, excess lube, kinks, and wires and spooled correctly			
Reeving correct			
Wedge sockets and wire rope clips not distorted, cracked, or missing			
Block not damaged			
Ball and hook is free to swivel and rotate			
Guards are in place			
Outrigger float(s) secured with pad pin			
Cab			
Handrails in place and not damaged			
Operator's manual in vehicle			
Load chart legible and visible to operator			
Hand signal chart visible to workers			
Charged fire extinguisher in place			
Cab glass not cracked and wipers are functional			

	Pass	Fail	N/A
Gauges and Indicators			
Load moment indicator operational			
Drum rotation indicator functioning			
Boom length indicator functioning			
Boom angle indicator functioning			
Engine: hydraulic, air, electrical, oil pressure, temperature, and fuel			
Operational Inspection			
Correct counterweight for the load			
Main hoist control functioning			
Auxiliary hoist control functioning			
Anti-two block in place and functioning			
Swing brake			
Lights and horns functional			

Notes:

Date_____

Operator
Crane number

Signature	
Model	

Visual Inspection	Pass	Fail	N/A
Engine fluid level correct (check dip stick or sight glass)			
Hydraulic fluid level correct (check dip stick or sight glass)			
Hydraulic system exhibits no apparent weeping or leaks			
Air system exhibits no audible leaks			
Tire pressure acceptable and tire not damaged			
Telescoping boom exhibits no damage to structure, wear pads, boom stops, or cylinder			
Wire rope free of dirt, excess lube, kinks, and wires and spooled correctly			
Reeving correct			
Wedge sockets and wire rope clips not distorted, cracked, or missing			
Block not damaged			
Ball and hook is free to swivel and rotate			
Guards are in place			
Outrigger float(s) secured with pad pin			
Cab			
Handrails in place and not damaged			
Operator's manual in vehicle			
Load chart legible and visible to operator			
Hand signal chart visible to workers			
Charged fire extinguisher in place			
Cab glass not cracked and wipers are functional			

	Pass	Fail	N/A
Gauges and Indicators			
Load moment indicator operational			
Drum rotation indicator functioning			
Boom length indicator functioning			
Boom angle indicator functioning			
Engine: hydraulic, air, electrical, oil pressure, temperature, and fuel			
Operational Inspection			
Correct counterweight for the load			
Main hoist control functioning			
Auxiliary hoist control functioning			
Anti-two block in place and functioning			
Swing brake			
Lights and horns functional			

Notes:

Date_____

Operator		Signature	
Crane number		Model	

Visual Inspection	Pass	Fail	N/A
Engine fluid level correct (check dip stick or sight glass)			
Hydraulic fluid level correct (check dip stick or sight glass)			
Hydraulic system exhibits no apparent weeping or leaks			
Air system exhibits no audible leaks			
Tire pressure acceptable and tire not damaged			
Telescoping boom exhibits no damage to structure, wear pads, boom stops, or cylinder			
Wire rope free of dirt, excess lube, kinks, and wires and spooled correctly			
Reeving correct			
Wedge sockets and wire rope clips not distorted, cracked, or missing			
Block not damaged			
Ball and hook is free to swivel and rotate			
Guards are in place			
Outrigger float(s) secured with pad pin			
Cab			
Handrails in place and not damaged			
Operator's manual in vehicle			
Load chart legible and visible to operator			
Hand signal chart visible to workers			
Charged fire extinguisher in place			
Cab glass not cracked and wipers are functional			

	Pass	Fail	N/A
Gauges and Indicators			
Load moment indicator operational			
Drum rotation indicator functioning			
Boom length indicator functioning			
Boom angle indicator functioning			
Engine: hydraulic, air, electrical, oil pressure, temperature, and fuel			
Operational Inspection			
Correct counterweight for the load			
Main hoist control functioning			
Auxiliary hoist control functioning			
Anti-two block in place and functioning			
Swing brake			
Lights and horns functional			

Notes:

Date_____

Operator
Crane number

Signature	
Model	

Visual Inspection	Pass	Fail	N/A
Engine fluid level correct (check dip stick or sight glass)			
Hydraulic fluid level correct (check dip stick or sight glass)			
Hydraulic system exhibits no apparent weeping or leaks			
Air system exhibits no audible leaks			
Tire pressure acceptable and tire not damaged			
Telescoping boom exhibits no damage to structure, wear pads, boom stops, or cylinder			
Wire rope free of dirt, excess lube, kinks, and wires and spooled correctly			
Reeving correct			
Wedge sockets and wire rope clips not distorted, cracked, or missing			
Block not damaged			
Ball and hook is free to swivel and rotate			
Guards are in place			
Outrigger float(s) secured with pad pin			
Cab			
Handrails in place and not damaged			
Operator's manual in vehicle			
Load chart legible and visible to operator			
Hand signal chart visible to workers			
Charged fire extinguisher in place			
Cab glass not cracked and wipers are functional			

	Pass	Fail	N/A
Gauges and Indicators			
Load moment indicator operational			
Drum rotation indicator functioning			
Boom length indicator functioning			
Boom angle indicator functioning			
Engine: hydraulic, air, electrical, oil pressure, temperature, and fuel			
Operational Inspection			
Correct counterweight for the load			
Main hoist control functioning			
Auxiliary hoist control functioning			
Anti-two block in place and functioning			
Swing brake			
Lights and horns functional			

Notes:

Date_____

Operator		Signature	
Crane number		Model	

Visual Inspection	Pass	Fail	N/A
Engine fluid level correct (check dip stick or sight glass)			
Hydraulic fluid level correct (check dip stick or sight glass)			
Hydraulic system exhibits no apparent weeping or leaks			
Air system exhibits no audible leaks			
Tire pressure acceptable and tire not damaged			
Telescoping boom exhibits no damage to structure, wear pads, boom stops, or cylinder			
Wire rope free of dirt, excess lube, kinks, and wires and spooled correctly			
Reeving correct			
Wedge sockets and wire rope clips not distorted, cracked, or missing			
Block not damaged			
Ball and hook is free to swivel and rotate			
Guards are in place			
Outrigger float(s) secured with pad pin			
Cab			
Handrails in place and not damaged			
Operator's manual in vehicle			
Load chart legible and visible to operator			
Hand signal chart visible to workers			
Charged fire extinguisher in place			
Cab glass not cracked and wipers are functional			

	Pass	Fail	N/A
Gauges and Indicators			
Load moment indicator operational			
Drum rotation indicator functioning			
Boom length indicator functioning			
Boom angle indicator functioning			
Engine: hydraulic, air, electrical, oil pressure, temperature, and fuel			
Operational Inspection			
Correct counterweight for the load			
Main hoist control functioning			
Auxiliary hoist control functioning			
Anti-two block in place and functioning			
Swing brake			
Lights and horns functional			

Notes:

Date_____

Operator
Crane number

Signature	
Model	

Visual Inspection	Pass	Fail	N/A
Engine fluid level correct (check dip stick or sight glass)			
Hydraulic fluid level correct (check dip stick or sight glass)			
Hydraulic system exhibits no apparent weeping or leaks			
Air system exhibits no audible leaks			
Tire pressure acceptable and tire not damaged			
Telescoping boom exhibits no damage to structure, wear pads, boom stops, or cylinder			
Wire rope free of dirt, excess lube, kinks, and wires and spooled correctly			
Reeving correct			
Wedge sockets and wire rope clips not distorted, cracked, or missing			
Block not damaged			
Ball and hook is free to swivel and rotate			
Guards are in place			
Outrigger float(s) secured with pad pin			
Cab			
Handrails in place and not damaged			
Operator's manual in vehicle			
Load chart legible and visible to operator			
Hand signal chart visible to workers			
Charged fire extinguisher in place			
Cab glass not cracked and wipers are functional			

	Pass	Fail	N/A
Gauges and Indicators			
Load moment indicator operational			
Drum rotation indicator functioning			
Boom length indicator functioning			
Boom angle indicator functioning			
Engine: hydraulic, air, electrical, oil pressure, temperature, and fuel			
Operational Inspection			
Correct counterweight for the load			
Main hoist control functioning			
Auxiliary hoist control functioning			
Anti-two block in place and functioning			
Swing brake			
Lights and horns functional			

Notes:

Date_____

Operator		Signature	
Crane number		Model	

Visual Inspection	Pass	Fail	N/A
Engine fluid level correct (check dip stick or sight glass)			
Hydraulic fluid level correct (check dip stick or sight glass)			
Hydraulic system exhibits no apparent weeping or leaks			
Air system exhibits no audible leaks			
Tire pressure acceptable and tire not damaged			
Telescoping boom exhibits no damage to structure, wear pads, boom stops, or cylinder			
Wire rope free of dirt, excess lube, kinks, and wires and spooled correctly			
Reeving correct			
Wedge sockets and wire rope clips not distorted, cracked, or missing			
Block not damaged			
Ball and hook is free to swivel and rotate			
Guards are in place			
Outrigger float(s) secured with pad pin			
Cab			
Handrails in place and not damaged			
Operator's manual in vehicle			
Load chart legible and visible to operator			
Hand signal chart visible to workers			
Charged fire extinguisher in place			
Cab glass not cracked and wipers are functional			

	Pass	Fail	N/A
Gauges and Indicators			
Load moment indicator operational			
Drum rotation indicator functioning			
Boom length indicator functioning			
Boom angle indicator functioning			
Engine: hydraulic, air, electrical, oil pressure, temperature, and fuel			
Operational Inspection			
Correct counterweight for the load			
Main hoist control functioning			
Auxiliary hoist control functioning			
Anti-two block in place and functioning			
Swing brake			
Lights and horns functional			

Notes:

Date_____

Operator	
Crane number	

Signature	
Model	

Visual Inspection	Pass	Fail	N/A
Engine fluid level correct (check dip stick or sight glass)			
Hydraulic fluid level correct (check dip stick or sight glass)			
Hydraulic system exhibits no apparent weeping or leaks			
Air system exhibits no audible leaks			
Tire pressure acceptable and tire not damaged			
Telescoping boom exhibits no damage to structure, wear pads, boom stops, or cylinder			
Wire rope free of dirt, excess lube, kinks, and wires and spooled correctly			
Reeving correct			
Wedge sockets and wire rope clips not distorted, cracked, or missing			
Block not damaged			
Ball and hook is free to swivel and rotate			
Guards are in place			
Outrigger float(s) secured with pad pin			
Cab			
Handrails in place and not damaged			
Operator's manual in vehicle			
Load chart legible and visible to operator			
Hand signal chart visible to workers			
Charged fire extinguisher in place			
Cab glass not cracked and wipers are functional			

	Pass	Fail	N/A
Gauges and Indicators			
Load moment indicator operational			
Drum rotation indicator functioning			
Boom length indicator functioning			
Boom angle indicator functioning			
Engine: hydraulic, air, electrical, oil pressure, temperature, and fuel			
Operational Inspection			
Correct counterweight for the load			
Main hoist control functioning			
Auxiliary hoist control functioning			
Anti-two block in place and functioning			
Swing brake			
Lights and horns functional			

Notes:

Date_____

Operator		Signature	
Crane number		Model	

Visual Inspection	Pass	Fail	N/A		Pass	Fail	N/A
Engine fluid level correct (check dip stick or sight glass)				**Gauges and Indicators**			
Hydraulic fluid level correct (check dip stick or sight glass)				Load moment indicator operational			
Hydraulic system exhibits no apparent weeping or leaks				Drum rotation indicator functioning			
Air system exhibits no audible leaks				Boom length indicator functioning			
Tire pressure acceptable and tire not damaged				Boom angle indicator functioning			
Telescoping boom exhibits no damage to structure, wear pads, boom stops, or cylinder				Engine: hydraulic, air, electrical, oil pressure, temperature, and fuel			
Wire rope free of dirt, excess lube, kinks, and wires and spooled correctly				**Operational Inspection**			
Reeving correct				Correct counterweight for the load			
Wedge sockets and wire rope clips not distorted, cracked, or missing				Main hoist control functioning			
Block not damaged				Auxiliary hoist control functioning			
Ball and hook is free to swivel and rotate				Anti-two block in place and functioning			
Guards are in place				Swing brake			
Outrigger float(s) secured with pad pin				Lights and horns functional			
Cab							
Handrails in place and not damaged							
Operator's manual in vehicle							
Load chart legible and visible to operator							
Hand signal chart visible to workers							
Charged fire extinguisher in place							
Cab glass not cracked and wipers are functional							

Notes:

Date_____

Operator		Signature	
Crane number		Model	

Visual Inspection	Pass	Fail	N/A
Engine fluid level correct (check dip stick or sight glass)			
Hydraulic fluid level correct (check dip stick or sight glass)			
Hydraulic system exhibits no apparent weeping or leaks			
Air system exhibits no audible leaks			
Tire pressure acceptable and tire not damaged			
Telescoping boom exhibits no damage to structure, wear pads, boom stops, or cylinder			
Wire rope free of dirt, excess lube, kinks, and wires and spooled correctly			
Reeving correct			
Wedge sockets and wire rope clips not distorted, cracked, or missing			
Block not damaged			
Ball and hook is free to swivel and rotate			
Guards are in place			
Outrigger float(s) secured with pad pin			
Cab			
Handrails in place and not damaged			
Operator's manual in vehicle			
Load chart legible and visible to operator			
Hand signal chart visible to workers			
Charged fire extinguisher in place			
Cab glass not cracked and wipers are functional			

	Pass	Fail	N/A
Gauges and Indicators			
Load moment indicator operational			
Drum rotation indicator functioning			
Boom length indicator functioning			
Boom angle indicator functioning			
Engine: hydraulic, air, electrical, oil pressure, temperature, and fuel			
Operational Inspection			
Correct counterweight for the load			
Main hoist control functioning			
Auxiliary hoist control functioning			
Anti-two block in place and functioning			
Swing brake			
Lights and horns functional			

Notes:

Date_____

Operator		Signature	
Crane number		Model	

Visual Inspection	Pass	Fail	N/A
Engine fluid level correct (check dip stick or sight glass)			
Hydraulic fluid level correct (check dip stick or sight glass)			
Hydraulic system exhibits no apparent weeping or leaks			
Air system exhibits no audible leaks			
Tire pressure acceptable and tire not damaged			
Telescoping boom exhibits no damage to structure, wear pads, boom stops, or cylinder			
Wire rope free of dirt, excess lube, kinks, and wires and spooled correctly			
Reeving correct			
Wedge sockets and wire rope clips not distorted, cracked, or missing			
Block not damaged			
Ball and hook is free to swivel and rotate			
Guards are in place			
Outrigger float(s) secured with pad pin			

Cab			
Handrails in place and not damaged			
Operator's manual in vehicle			
Load chart legible and visible to operator			
Hand signal chart visible to workers			
Charged fire extinguisher in place			
Cab glass not cracked and wipers are functional			

Gauges and Indicators	Pass	Fail	N/A
Load moment indicator operational			
Drum rotation indicator functioning			
Boom length indicator functioning			
Boom angle indicator functioning			
Engine: hydraulic, air, electrical, oil pressure, temperature, and fuel			

Operational Inspection			
Correct counterweight for the load			
Main hoist control functioning			
Auxiliary hoist control functioning			
Anti-two block in place and functioning			
Swing brake			
Lights and horns functional			

Notes:

Date_____

Operator	
Crane number	

Signature		
Model		

Visual Inspection	Pass	Fail	N/A
Engine fluid level correct (check dip stick or sight glass)			
Hydraulic fluid level correct (check dip stick or sight glass)			
Hydraulic system exhibits no apparent weeping or leaks			
Air system exhibits no audible leaks			
Tire pressure acceptable and tire not damaged			
Telescoping boom exhibits no damage to structure, wear pads, boom stops, or cylinder			
Wire rope free of dirt, excess lube, kinks, and wires and spooled correctly			
Reeving correct			
Wedge sockets and wire rope clips not distorted, cracked, or missing			
Block not damaged			
Ball and hook is free to swivel and rotate			
Guards are in place			
Outrigger float(s) secured with pad pin			
Cab			
Handrails in place and not damaged			
Operator's manual in vehicle			
Load chart legible and visible to operator			
Hand signal chart visible to workers			
Charged fire extinguisher in place			
Cab glass not cracked and wipers are functional			

	Pass	Fail	N/A
Gauges and Indicators			
Load moment indicator operational			
Drum rotation indicator functioning			
Boom length indicator functioning			
Boom angle indicator functioning			
Engine: hydraulic, air, electrical, oil pressure, temperature, and fuel			
Operational Inspection			
Correct counterweight for the load			
Main hoist control functioning			
Auxiliary hoist control functioning			
Anti-two block in place and functioning			
Swing brake			
Lights and horns functional			

Notes:

Date_____

Operator		Signature	
Crane number		Model	

Visual Inspection	Pass	Fail	N/A
Engine fluid level correct (check dip stick or sight glass)			
Hydraulic fluid level correct (check dip stick or sight glass)			
Hydraulic system exhibits no apparent weeping or leaks			
Air system exhibits no audible leaks			
Tire pressure acceptable and tire not damaged			
Telescoping boom exhibits no damage to structure, wear pads, boom stops, or cylinder			
Wire rope free of dirt, excess lube, kinks, and wires and spooled correctly			
Reeving correct			
Wedge sockets and wire rope clips not distorted, cracked, or missing			
Block not damaged			
Ball and hook is free to swivel and rotate			
Guards are in place			
Outrigger float(s) secured with pad pin			
Cab			
Handrails in place and not damaged			
Operator's manual in vehicle			
Load chart legible and visible to operator			
Hand signal chart visible to workers			
Charged fire extinguisher in place			
Cab glass not cracked and wipers are functional			

	Pass	Fail	N/A
Gauges and Indicators			
Load moment indicator operational			
Drum rotation indicator functioning			
Boom length indicator functioning			
Boom angle indicator functioning			
Engine: hydraulic, air, electrical, oil pressure, temperature, and fuel			
Operational Inspection			
Correct counterweight for the load			
Main hoist control functioning			
Auxiliary hoist control functioning			
Anti-two block in place and functioning			
Swing brake			
Lights and horns functional			

Notes:

Date _____

Operator
Crane number

Signature	
Model	

Visual Inspection	Pass	Fail	N/A
Engine fluid level correct (check dip stick or sight glass)			
Hydraulic fluid level correct (check dip stick or sight glass)			
Hydraulic system exhibits no apparent weeping or leaks			
Air system exhibits no audible leaks			
Tire pressure acceptable and tire not damaged			
Telescoping boom exhibits no damage to structure, wear pads, boom stops, or cylinder			
Wire rope free of dirt, excess lube, kinks, and wires and spooled correctly			
Reeving correct			
Wedge sockets and wire rope clips not distorted, cracked, or missing			
Block not damaged			
Ball and hook is free to swivel and rotate			
Guards are in place			
Outrigger float(s) secured with pad pin			
Cab			
Handrails in place and not damaged			
Operator's manual in vehicle			
Load chart legible and visible to operator			
Hand signal chart visible to workers			
Charged fire extinguisher in place			
Cab glass not cracked and wipers are functional			

	Pass	Fail	N/A
Gauges and Indicators			
Load moment indicator operational			
Drum rotation indicator functioning			
Boom length indicator functioning			
Boom angle indicator functioning			
Engine: hydraulic, air, electrical, oil pressure, temperature, and fuel			
Operational Inspection			
Correct counterweight for the load			
Main hoist control functioning			
Auxiliary hoist control functioning			
Anti-two block in place and functioning			
Swing brake			
Lights and horns functional			

Notes:

Date_____

Operator		Signature	
Crane number		Model	

Visual Inspection	Pass	Fail	N/A		Pass	Fail	N/A
Engine fluid level correct (check dip stick or sight glass)				**Gauges and Indicators**			
Hydraulic fluid level correct (check dip stick or sight glass)				Load moment indicator operational			
Hydraulic system exhibits no apparent weeping or leaks				Drum rotation indicator functioning			
Air system exhibits no audible leaks				Boom length indicator functioning			
Tire pressure acceptable and tire not damaged				Boom angle indicator functioning			
Telescoping boom exhibits no damage to structure, wear pads, boom stops, or cylinder				Engine: hydraulic, air, electrical, oil pressure, temperature, and fuel			
Wire rope free of dirt, excess lube, kinks, and wires and spooled correctly				**Operational Inspection**			
Reeving correct				Correct counterweight for the load			
Wedge sockets and wire rope clips not distorted, cracked, or missing				Main hoist control functioning			
Block not damaged				Auxiliary hoist control functioning			
Ball and hook is free to swivel and rotate				Anti-two block in place and functioning			
Guards are in place				Swing brake			
Outrigger float(s) secured with pad pin				Lights and horns functional			
Cab							
Handrails in place and not damaged							
Operator's manual in vehicle							
Load chart legible and visible to operator							
Hand signal chart visible to workers							
Charged fire extinguisher in place							
Cab glass not cracked and wipers are functional							

Notes:

Date_____

Operator		Signature	
Crane number		Model	

Visual Inspection	Pass	Fail	N/A
Engine fluid level correct (check dip stick or sight glass)			
Hydraulic fluid level correct (check dip stick or sight glass)			
Hydraulic system exhibits no apparent weeping or leaks			
Air system exhibits no audible leaks			
Tire pressure acceptable and tire not damaged			
Telescoping boom exhibits no damage to structure, wear pads, boom stops, or cylinder			
Wire rope free of dirt, excess lube, kinks, and wires and spooled correctly			
Reeving correct			
Wedge sockets and wire rope clips not distorted, cracked, or missing			
Block not damaged			
Ball and hook is free to swivel and rotate			
Guards are in place			
Outrigger float(s) secured with pad pin			
Cab			
Handrails in place and not damaged			
Operator's manual in vehicle			
Load chart legible and visible to operator			
Hand signal chart visible to workers			
Charged fire extinguisher in place			
Cab glass not cracked and wipers are functional			

	Pass	Fail	N/A
Gauges and Indicators			
Load moment indicator operational			
Drum rotation indicator functioning			
Boom length indicator functioning			
Boom angle indicator functioning			
Engine: hydraulic, air, electrical, oil pressure, temperature, and fuel			
Operational Inspection			
Correct counterweight for the load			
Main hoist control functioning			
Auxiliary hoist control functioning			
Anti-two block in place and functioning			
Swing brake			
Lights and horns functional			

Notes:

Date_____

Operator		Signature	
Crane number		Model	

Visual Inspection	Pass	Fail	N/A		Pass	Fail	N/A
Engine fluid level correct (check dip stick or sight glass)				**Gauges and Indicators**			
Hydraulic fluid level correct (check dip stick or sight glass)				Load moment indicator operational			
Hydraulic system exhibits no apparent weeping or leaks				Drum rotation indicator functioning			
Air system exhibits no audible leaks				Boom length indicator functioning			
Tire pressure acceptable and tire not damaged				Boom angle indicator functioning			
Telescoping boom exhibits no damage to structure, wear pads, boom stops, or cylinder				Engine: hydraulic, air, electrical, oil pressure, temperature, and fuel			
Wire rope free of dirt, excess lube, kinks, and wires and spooled correctly				**Operational Inspection**			
Reeving correct				Correct counterweight for the load			
Wedge sockets and wire rope clips not distorted, cracked, or missing				Main hoist control functioning			
Block not damaged				Auxiliary hoist control functioning			
Ball and hook is free to swivel and rotate				Anti-two block in place and functioning			
Guards are in place				Swing brake			
Outrigger float(s) secured with pad pin				Lights and horns functional			
Cab							
Handrails in place and not damaged							
Operator's manual in vehicle							
Load chart legible and visible to operator							
Hand signal chart visible to workers							
Charged fire extinguisher in place							
Cab glass not cracked and wipers are functional							

Notes:

Date_____

Operator
Crane number

Signature	
Model	

Visual Inspection	Pass	Fail	N/A
Engine fluid level correct (check dip stick or sight glass)			
Hydraulic fluid level correct (check dip stick or sight glass)			
Hydraulic system exhibits no apparent weeping or leaks			
Air system exhibits no audible leaks			
Tire pressure acceptable and tire not damaged			
Telescoping boom exhibits no damage to structure, wear pads, boom stops, or cylinder			
Wire rope free of dirt, excess lube, kinks, and wires and spooled correctly			
Reeving correct			
Wedge sockets and wire rope clips not distorted, cracked, or missing			
Block not damaged			
Ball and hook is free to swivel and rotate			
Guards are in place			
Outrigger float(s) secured with pad pin			
Cab			
Handrails in place and not damaged			
Operator's manual in vehicle			
Load chart legible and visible to operator			
Hand signal chart visible to workers			
Charged fire extinguisher in place			
Cab glass not cracked and wipers are functional			

	Pass	Fail	N/A
Gauges and Indicators			
Load moment indicator operational			
Drum rotation indicator functioning			
Boom length indicator functioning			
Boom angle indicator functioning			
Engine: hydraulic, air, electrical, oil pressure, temperature, and fuel			
Operational Inspection			
Correct counterweight for the load			
Main hoist control functioning			
Auxiliary hoist control functioning			
Anti-two block in place and functioning			
Swing brake			
Lights and horns functional			

Notes:

Date_____

Operator		Signature	
Crane number		Model	

Visual Inspection	Pass	Fail	N/A		Pass	Fail	N/A
Engine fluid level correct (check dip stick or sight glass)				**Gauges and Indicators**			
Hydraulic fluid level correct (check dip stick or sight glass)				Load moment indicator operational			
Hydraulic system exhibits no apparent weeping or leaks				Drum rotation indicator functioning			
Air system exhibits no audible leaks				Boom length indicator functioning			
Tire pressure acceptable and tire not damaged				Boom angle indicator functioning			
Telescoping boom exhibits no damage to structure, wear pads, boom stops, or cylinder				Engine: hydraulic, air, electrical, oil pressure, temperature, and fuel			
Wire rope free of dirt, excess lube, kinks, and wires and spooled correctly				**Operational Inspection**			
Reeving correct				Correct counterweight for the load			
Wedge sockets and wire rope clips not distorted, cracked, or missing				Main hoist control functioning			
Block not damaged				Auxiliary hoist control functioning			
Ball and hook is free to swivel and rotate				Anti-two block in place and functioning			
Guards are in place				Swing brake			
Outrigger float(s) secured with pad pin				Lights and horns functional			
Cab							
Handrails in place and not damaged							
Operator's manual in vehicle							
Load chart legible and visible to operator							
Hand signal chart visible to workers							
Charged fire extinguisher in place							
Cab glass not cracked and wipers are functional							

Notes:

Date_____

Operator
Crane number

Signature	
Model	

Visual Inspection	Pass	Fail	N/A
Engine fluid level correct (check dip stick or sight glass)			
Hydraulic fluid level correct (check dip stick or sight glass)			
Hydraulic system exhibits no apparent weeping or leaks			
Air system exhibits no audible leaks			
Tire pressure acceptable and tire not damaged			
Telescoping boom exhibits no damage to structure, wear pads, boom stops, or cylinder			
Wire rope free of dirt, excess lube, kinks, and wires and spooled correctly			
Reeving correct			
Wedge sockets and wire rope clips not distorted, cracked, or missing			
Block not damaged			
Ball and hook is free to swivel and rotate			
Guards are in place			
Outrigger float(s) secured with pad pin			
Cab			
Handrails in place and not damaged			
Operator's manual in vehicle			
Load chart legible and visible to operator			
Hand signal chart visible to workers			
Charged fire extinguisher in place			
Cab glass not cracked and wipers are functional			

	Pass	Fail	N/A
Gauges and Indicators			
Load moment indicator operational			
Drum rotation indicator functioning			
Boom length indicator functioning			
Boom angle indicator functioning			
Engine: hydraulic, air, electrical, oil pressure, temperature, and fuel			
Operational Inspection			
Correct counterweight for the load			
Main hoist control functioning			
Auxiliary hoist control functioning			
Anti-two block in place and functioning			
Swing brake			
Lights and horns functional			

Notes:

Date_____

Operator	
Crane number	

Signature	
Model	

Visual Inspection	Pass	Fail	N/A
Engine fluid level correct (check dip stick or sight glass)			
Hydraulic fluid level correct (check dip stick or sight glass)			
Hydraulic system exhibits no apparent weeping or leaks			
Air system exhibits no audible leaks			
Tire pressure acceptable and tire not damaged			
Telescoping boom exhibits no damage to structure, wear pads, boom stops, or cylinder			
Wire rope free of dirt, excess lube, kinks, and wires and spooled correctly			
Reeving correct			
Wedge sockets and wire rope clips not distorted, cracked, or missing			
Block not damaged			
Ball and hook is free to swivel and rotate			
Guards are in place			
Outrigger float(s) secured with pad pin			
Cab			
Handrails in place and not damaged			
Operator's manual in vehicle			
Load chart legible and visible to operator			
Hand signal chart visible to workers			
Charged fire extinguisher in place			
Cab glass not cracked and wipers are functional			

	Pass	Fail	N/A
Gauges and Indicators			
Load moment indicator operational			
Drum rotation indicator functioning			
Boom length indicator functioning			
Boom angle indicator functioning			
Engine: hydraulic, air, electrical, oil pressure, temperature, and fuel			
Operational Inspection			
Correct counterweight for the load			
Main hoist control functioning			
Auxiliary hoist control functioning			
Anti-two block in place and functioning			
Swing brake			
Lights and horns functional			

Notes:

Date_____

Operator
Crane number

Signature	
Model	

Visual Inspection	Pass	Fail	N/A
Engine fluid level correct (check dip stick or sight glass)			
Hydraulic fluid level correct (check dip stick or sight glass)			
Hydraulic system exhibits no apparent weeping or leaks			
Air system exhibits no audible leaks			
Tire pressure acceptable and tire not damaged			
Telescoping boom exhibits no damage to structure, wear pads, boom stops, or cylinder			
Wire rope free of dirt, excess lube, kinks, and wires and spooled correctly			
Reeving correct			
Wedge sockets and wire rope clips not distorted, cracked, or missing			
Block not damaged			
Ball and hook is free to swivel and rotate			
Guards are in place			
Outrigger float(s) secured with pad pin			
Cab			
Handrails in place and not damaged			
Operator's manual in vehicle			
Load chart legible and visible to operator			
Hand signal chart visible to workers			
Charged fire extinguisher in place			
Cab glass not cracked and wipers are functional			

	Pass	Fail	N/A
Gauges and Indicators			
Load moment indicator operational			
Drum rotation indicator functioning			
Boom length indicator functioning			
Boom angle indicator functioning			
Engine: hydraulic, air, electrical, oil pressure, temperature, and fuel			
Operational Inspection			
Correct counterweight for the load			
Main hoist control functioning			
Auxiliary hoist control functioning			
Anti-two block in place and functioning			
Swing brake			
Lights and horns functional			

Notes:

Date_____

Operator		Signature	
Crane number		Model	

Visual Inspection	Pass	Fail	N/A
Engine fluid level correct (check dip stick or sight glass)			
Hydraulic fluid level correct (check dip stick or sight glass)			
Hydraulic system exhibits no apparent weeping or leaks			
Air system exhibits no audible leaks			
Tire pressure acceptable and tire not damaged			
Telescoping boom exhibits no damage to structure, wear pads, boom stops, or cylinder			
Wire rope free of dirt, excess lube, kinks, and wires and spooled correctly			
Reeving correct			
Wedge sockets and wire rope clips not distorted, cracked, or missing			
Block not damaged			
Ball and hook is free to swivel and rotate			
Guards are in place			
Outrigger float(s) secured with pad pin			
Cab			
Handrails in place and not damaged			
Operator's manual in vehicle			
Load chart legible and visible to operator			
Hand signal chart visible to workers			
Charged fire extinguisher in place			
Cab glass not cracked and wipers are functional			

	Pass	Fail	N/A
Gauges and Indicators			
Load moment indicator operational			
Drum rotation indicator functioning			
Boom length indicator functioning			
Boom angle indicator functioning			
Engine: hydraulic, air, electrical, oil pressure, temperature, and fuel			
Operational Inspection			
Correct counterweight for the load			
Main hoist control functioning			
Auxiliary hoist control functioning			
Anti-two block in place and functioning			
Swing brake			
Lights and horns functional			

Notes:

Date_____

Operator
Crane number

Signature	
Model	

Visual Inspection	Pass	Fail	N/A
Engine fluid level correct (check dip stick or sight glass)			
Hydraulic fluid level correct (check dip stick or sight glass)			
Hydraulic system exhibits no apparent weeping or leaks			
Air system exhibits no audible leaks			
Tire pressure acceptable and tire not damaged			
Telescoping boom exhibits no damage to structure, wear pads, boom stops, or cylinder			
Wire rope free of dirt, excess lube, kinks, and wires and spooled correctly			
Reeving correct			
Wedge sockets and wire rope clips not distorted, cracked, or missing			
Block not damaged			
Ball and hook is free to swivel and rotate			
Guards are in place			
Outrigger float(s) secured with pad pin			
Cab			
Handrails in place and not damaged			
Operator's manual in vehicle			
Load chart legible and visible to operator			
Hand signal chart visible to workers			
Charged fire extinguisher in place			
Cab glass not cracked and wipers are functional			

	Pass	Fail	N/A
Gauges and Indicators			
Load moment indicator operational			
Drum rotation indicator functioning			
Boom length indicator functioning			
Boom angle indicator functioning			
Engine: hydraulic, air, electrical, oil pressure, temperature, and fuel			
Operational Inspection			
Correct counterweight for the load			
Main hoist control functioning			
Auxiliary hoist control functioning			
Anti-two block in place and functioning			
Swing brake			
Lights and horns functional			

Notes:

Date_____

Operator		Signature	
Crane number		Model	

Visual Inspection	Pass	Fail	N/A
Engine fluid level correct (check dip stick or sight glass)			
Hydraulic fluid level correct (check dip stick or sight glass)			
Hydraulic system exhibits no apparent weeping or leaks			
Air system exhibits no audible leaks			
Tire pressure acceptable and tire not damaged			
Telescoping boom exhibits no damage to structure, wear pads, boom stops, or cylinder			
Wire rope free of dirt, excess lube, kinks, and wires and spooled correctly			
Reeving correct			
Wedge sockets and wire rope clips not distorted, cracked, or missing			
Block not damaged			
Ball and hook is free to swivel and rotate			
Guards are in place			
Outrigger float(s) secured with pad pin			
Cab			
Handrails in place and not damaged			
Operator's manual in vehicle			
Load chart legible and visible to operator			
Hand signal chart visible to workers			
Charged fire extinguisher in place			
Cab glass not cracked and wipers are functional			

	Pass	Fail	N/A
Gauges and Indicators			
Load moment indicator operational			
Drum rotation indicator functioning			
Boom length indicator functioning			
Boom angle indicator functioning			
Engine: hydraulic, air, electrical, oil pressure, temperature, and fuel			
Operational Inspection			
Correct counterweight for the load			
Main hoist control functioning			
Auxiliary hoist control functioning			
Anti-two block in place and functioning			
Swing brake			
Lights and horns functional			

Notes:

Date_____

Operator
Crane number

Signature	
Model	

Visual Inspection	Pass	Fail	N/A
Engine fluid level correct (check dip stick or sight glass)			
Hydraulic fluid level correct (check dip stick or sight glass)			
Hydraulic system exhibits no apparent weeping or leaks			
Air system exhibits no audible leaks			
Tire pressure acceptable and tire not damaged			
Telescoping boom exhibits no damage to structure, wear pads, boom stops, or cylinder			
Wire rope free of dirt, excess lube, kinks, and wires and spooled correctly			
Reeving correct			
Wedge sockets and wire rope clips not distorted, cracked, or missing			
Block not damaged			
Ball and hook is free to swivel and rotate			
Guards are in place			
Outrigger float(s) secured with pad pin			
Cab			
Handrails in place and not damaged			
Operator's manual in vehicle			
Load chart legible and visible to operator			
Hand signal chart visible to workers			
Charged fire extinguisher in place			
Cab glass not cracked and wipers are functional			

	Pass	Fail	N/A
Gauges and Indicators			
Load moment indicator operational			
Drum rotation indicator functioning			
Boom length indicator functioning			
Boom angle indicator functioning			
Engine: hydraulic, air, electrical, oil pressure, temperature, and fuel			
Operational Inspection			
Correct counterweight for the load			
Main hoist control functioning			
Auxiliary hoist control functioning			
Anti-two block in place and functioning			
Swing brake			
Lights and horns functional			

Notes:

Date_____

Operator		Signature	
Crane number		Model	

Visual Inspection	Pass	Fail	N/A
Engine fluid level correct (check dip stick or sight glass)			
Hydraulic fluid level correct (check dip stick or sight glass)			
Hydraulic system exhibits no apparent weeping or leaks			
Air system exhibits no audible leaks			
Tire pressure acceptable and tire not damaged			
Telescoping boom exhibits no damage to structure, wear pads, boom stops, or cylinder			
Wire rope free of dirt, excess lube, kinks, and wires and spooled correctly			
Reeving correct			
Wedge sockets and wire rope clips not distorted, cracked, or missing			
Block not damaged			
Ball and hook is free to swivel and rotate			
Guards are in place			
Outrigger float(s) secured with pad pin			
Cab			
Handrails in place and not damaged			
Operator's manual in vehicle			
Load chart legible and visible to operator			
Hand signal chart visible to workers			
Charged fire extinguisher in place			
Cab glass not cracked and wipers are functional			

	Pass	Fail	N/A
Gauges and Indicators			
Load moment indicator operational			
Drum rotation indicator functioning			
Boom length indicator functioning			
Boom angle indicator functioning			
Engine: hydraulic, air, electrical, oil pressure, temperature, and fuel			
Operational Inspection			
Correct counterweight for the load			
Main hoist control functioning			
Auxiliary hoist control functioning			
Anti-two block in place and functioning			
Swing brake			
Lights and horns functional			

Notes:

Date _____

Operator	
Crane number	

Signature		
Model		

Visual Inspection	Pass	Fail	N/A
Engine fluid level correct (check dip stick or sight glass)			
Hydraulic fluid level correct (check dip stick or sight glass)			
Hydraulic system exhibits no apparent weeping or leaks			
Air system exhibits no audible leaks			
Tire pressure acceptable and tire not damaged			
Telescoping boom exhibits no damage to structure, wear pads, boom stops, or cylinder			
Wire rope free of dirt, excess lube, kinks, and wires and spooled correctly			
Reeving correct			
Wedge sockets and wire rope clips not distorted, cracked, or missing			
Block not damaged			
Ball and hook is free to swivel and rotate			
Guards are in place			
Outrigger float(s) secured with pad pin			
Cab			
Handrails in place and not damaged			
Operator's manual in vehicle			
Load chart legible and visible to operator			
Hand signal chart visible to workers			
Charged fire extinguisher in place			
Cab glass not cracked and wipers are functional			

	Pass	Fail	N/A
Gauges and Indicators			
Load moment indicator operational			
Drum rotation indicator functioning			
Boom length indicator functioning			
Boom angle indicator functioning			
Engine: hydraulic, air, electrical, oil pressure, temperature, and fuel			
Operational Inspection			
Correct counterweight for the load			
Main hoist control functioning			
Auxiliary hoist control functioning			
Anti-two block in place and functioning			
Swing brake			
Lights and horns functional			

Notes:

Date_____

Operator		Signature	
Crane number		Model	

Visual Inspection	Pass	Fail	N/A
Engine fluid level correct (check dip stick or sight glass)			
Hydraulic fluid level correct (check dip stick or sight glass)			
Hydraulic system exhibits no apparent weeping or leaks			
Air system exhibits no audible leaks			
Tire pressure acceptable and tire not damaged			
Telescoping boom exhibits no damage to structure, wear pads, boom stops, or cylinder			
Wire rope free of dirt, excess lube, kinks, and wires and spooled correctly			
Reeving correct			
Wedge sockets and wire rope clips not distorted, cracked, or missing			
Block not damaged			
Ball and hook is free to swivel and rotate			
Guards are in place			
Outrigger float(s) secured with pad pin			
Cab			
Handrails in place and not damaged			
Operator's manual in vehicle			
Load chart legible and visible to operator			
Hand signal chart visible to workers			
Charged fire extinguisher in place			
Cab glass not cracked and wipers are functional			

	Pass	Fail	N/A
Gauges and Indicators			
Load moment indicator operational			
Drum rotation indicator functioning			
Boom length indicator functioning			
Boom angle indicator functioning			
Engine: hydraulic, air, electrical, oil pressure, temperature, and fuel			
Operational Inspection			
Correct counterweight for the load			
Main hoist control functioning			
Auxiliary hoist control functioning			
Anti-two block in place and functioning			
Swing brake			
Lights and horns functional			

Notes:

Date_____

Operator
Crane number

Signature	
Model	

Visual Inspection	Pass	Fail	N/A
Engine fluid level correct (check dip stick or sight glass)			
Hydraulic fluid level correct (check dip stick or sight glass)			
Hydraulic system exhibits no apparent weeping or leaks			
Air system exhibits no audible leaks			
Tire pressure acceptable and tire not damaged			
Telescoping boom exhibits no damage to structure, wear pads, boom stops, or cylinder			
Wire rope free of dirt, excess lube, kinks, and wires and spooled correctly			
Reeving correct			
Wedge sockets and wire rope clips not distorted, cracked, or missing			
Block not damaged			
Ball and hook is free to swivel and rotate			
Guards are in place			
Outrigger float(s) secured with pad pin			
Cab			
Handrails in place and not damaged			
Operator's manual in vehicle			
Load chart legible and visible to operator			
Hand signal chart visible to workers			
Charged fire extinguisher in place			
Cab glass not cracked and wipers are functional			

	Pass	Fail	N/A
Gauges and Indicators			
Load moment indicator operational			
Drum rotation indicator functioning			
Boom length indicator functioning			
Boom angle indicator functioning			
Engine: hydraulic, air, electrical, oil pressure, temperature, and fuel			
Operational Inspection			
Correct counterweight for the load			
Main hoist control functioning			
Auxiliary hoist control functioning			
Anti-two block in place and functioning			
Swing brake			
Lights and horns functional			

Notes:

Date _____

Operator		Signature	
Crane number		Model	

Visual Inspection	Pass	Fail	N/A
Engine fluid level correct (check dip stick or sight glass)			
Hydraulic fluid level correct (check dip stick or sight glass)			
Hydraulic system exhibits no apparent weeping or leaks			
Air system exhibits no audible leaks			
Tire pressure acceptable and tire not damaged			
Telescoping boom exhibits no damage to structure, wear pads, boom stops, or cylinder			
Wire rope free of dirt, excess lube, kinks, and wires and spooled correctly			
Reeving correct			
Wedge sockets and wire rope clips not distorted, cracked, or missing			
Block not damaged			
Ball and hook is free to swivel and rotate			
Guards are in place			
Outrigger float(s) secured with pad pin			
Cab			
Handrails in place and not damaged			
Operator's manual in vehicle			
Load chart legible and visible to operator			
Hand signal chart visible to workers			
Charged fire extinguisher in place			
Cab glass not cracked and wipers are functional			

	Pass	Fail	N/A
Gauges and Indicators			
Load moment indicator operational			
Drum rotation indicator functioning			
Boom length indicator functioning			
Boom angle indicator functioning			
Engine: hydraulic, air, electrical, oil pressure, temperature, and fuel			
Operational Inspection			
Correct counterweight for the load			
Main hoist control functioning			
Auxiliary hoist control functioning			
Anti-two block in place and functioning			
Swing brake			
Lights and horns functional			

Notes:

Date_____

Operator
Crane number

Signature	
Model	

Visual Inspection	Pass	Fail	N/A
Engine fluid level correct (check dip stick or sight glass)			
Hydraulic fluid level correct (check dip stick or sight glass)			
Hydraulic system exhibits no apparent weeping or leaks			
Air system exhibits no audible leaks			
Tire pressure acceptable and tire not damaged			
Telescoping boom exhibits no damage to structure, wear pads, boom stops, or cylinder			
Wire rope free of dirt, excess lube, kinks, and wires and spooled correctly			
Reeving correct			
Wedge sockets and wire rope clips not distorted, cracked, or missing			
Block not damaged			
Ball and hook is free to swivel and rotate			
Guards are in place			
Outrigger float(s) secured with pad pin			
Cab			
Handrails in place and not damaged			
Operator's manual in vehicle			
Load chart legible and visible to operator			
Hand signal chart visible to workers			
Charged fire extinguisher in place			
Cab glass not cracked and wipers are functional			

	Pass	Fail	N/A
Gauges and Indicators			
Load moment indicator operational			
Drum rotation indicator functioning			
Boom length indicator functioning			
Boom angle indicator functioning			
Engine: hydraulic, air, electrical, oil pressure, temperature, and fuel			
Operational Inspection			
Correct counterweight for the load			
Main hoist control functioning			
Auxiliary hoist control functioning			
Anti-two block in place and functioning			
Swing brake			
Lights and horns functional			

Notes:

Date_____

Operator		Signature	
Crane number		Model	

Visual Inspection	Pass	Fail	N/A		Pass	Fail	N/A
Engine fluid level correct (check dip stick or sight glass)				**Gauges and Indicators**			
Hydraulic fluid level correct (check dip stick or sight glass)				Load moment indicator operational			
Hydraulic system exhibits no apparent weeping or leaks				Drum rotation indicator functioning			
Air system exhibits no audible leaks				Boom length indicator functioning			
Tire pressure acceptable and tire not damaged				Boom angle indicator functioning			
Telescoping boom exhibits no damage to structure, wear pads, boom stops, or cylinder				Engine: hydraulic, air, electrical, oil pressure, temperature, and fuel			
Wire rope free of dirt, excess lube, kinks, and wires and spooled correctly				**Operational Inspection**			
Reeving correct				Correct counterweight for the load			
Wedge sockets and wire rope clips not distorted, cracked, or missing				Main hoist control functioning			
Block not damaged				Auxiliary hoist control functioning			
Ball and hook is free to swivel and rotate				Anti-two block in place and functioning			
Guards are in place				Swing brake			
Outrigger float(s) secured with pad pin				Lights and horns functional			
Cab							
Handrails in place and not damaged							
Operator's manual in vehicle							
Load chart legible and visible to operator							
Hand signal chart visible to workers							
Charged fire extinguisher in place							
Cab glass not cracked and wipers are functional							

Notes:

Date_____

Operator	
Crane number	

Signature	
Model	

Visual Inspection	Pass	Fail	N/A
Engine fluid level correct (check dip stick or sight glass)			
Hydraulic fluid level correct (check dip stick or sight glass)			
Hydraulic system exhibits no apparent weeping or leaks			
Air system exhibits no audible leaks			
Tire pressure acceptable and tire not damaged			
Telescoping boom exhibits no damage to structure, wear pads, boom stops, or cylinder			
Wire rope free of dirt, excess lube, kinks, and wires and spooled correctly			
Reeving correct			
Wedge sockets and wire rope clips not distorted, cracked, or missing			
Block not damaged			
Ball and hook is free to swivel and rotate			
Guards are in place			
Outrigger float(s) secured with pad pin			
Cab			
Handrails in place and not damaged			
Operator's manual in vehicle			
Load chart legible and visible to operator			
Hand signal chart visible to workers			
Charged fire extinguisher in place			
Cab glass not cracked and wipers are functional			

	Pass	Fail	N/A
Gauges and Indicators			
Load moment indicator operational			
Drum rotation indicator functioning			
Boom length indicator functioning			
Boom angle indicator functioning			
Engine: hydraulic, air, electrical, oil pressure, temperature, and fuel			
Operational Inspection			
Correct counterweight for the load			
Main hoist control functioning			
Auxiliary hoist control functioning			
Anti-two block in place and functioning			
Swing brake			
Lights and horns functional			

Notes:

Date_____

Operator		Signature	
Crane number		Model	

Visual Inspection	Pass	Fail	N/A		Pass	Fail	N/A
Engine fluid level correct (check dip stick or sight glass)				**Gauges and Indicators**			
Hydraulic fluid level correct (check dip stick or sight glass)				Load moment indicator operational			
Hydraulic system exhibits no apparent weeping or leaks				Drum rotation indicator functioning			
Air system exhibits no audible leaks				Boom length indicator functioning			
Tire pressure acceptable and tire not damaged				Boom angle indicator functioning			
Telescoping boom exhibits no damage to structure, wear pads, boom stops, or cylinder				Engine: hydraulic, air, electrical, oil pressure, temperature, and fuel			
Wire rope free of dirt, excess lube, kinks, and wires and spooled correctly				**Operational Inspection**			
Reeving correct				Correct counterweight for the load			
Wedge sockets and wire rope clips not distorted, cracked, or missing				Main hoist control functioning			
Block not damaged				Auxiliary hoist control functioning			
Ball and hook is free to swivel and rotate				Anti-two block in place and functioning			
Guards are in place				Swing brake			
Outrigger float(s) secured with pad pin				Lights and horns functional			
Cab							
Handrails in place and not damaged							
Operator's manual in vehicle							
Load chart legible and visible to operator							
Hand signal chart visible to workers							
Charged fire extinguisher in place							
Cab glass not cracked and wipers are functional							

Notes:

Date_____

Operator
Crane number

Signature	
Model	

Visual Inspection	Pass	Fail	N/A
Engine fluid level correct (check dip stick or sight glass)			
Hydraulic fluid level correct (check dip stick or sight glass)			
Hydraulic system exhibits no apparent weeping or leaks			
Air system exhibits no audible leaks			
Tire pressure acceptable and tire not damaged			
Telescoping boom exhibits no damage to structure, wear pads, boom stops, or cylinder			
Wire rope free of dirt, excess lube, kinks, and wires and spooled correctly			
Reeving correct			
Wedge sockets and wire rope clips not distorted, cracked, or missing			
Block not damaged			
Ball and hook is free to swivel and rotate			
Guards are in place			
Outrigger float(s) secured with pad pin			
Cab			
Handrails in place and not damaged			
Operator's manual in vehicle			
Load chart legible and visible to operator			
Hand signal chart visible to workers			
Charged fire extinguisher in place			
Cab glass not cracked and wipers are functional			

	Pass	Fail	N/A
Gauges and Indicators			
Load moment indicator operational			
Drum rotation indicator functioning			
Boom length indicator functioning			
Boom angle indicator functioning			
Engine: hydraulic, air, electrical, oil pressure, temperature, and fuel			
Operational Inspection			
Correct counterweight for the load			
Main hoist control functioning			
Auxiliary hoist control functioning			
Anti-two block in place and functioning			
Swing brake			
Lights and horns functional			

Notes:

Date_____

Operator		Signature	
Crane number		Model	

Visual Inspection	Pass	Fail	N/A
Engine fluid level correct (check dip stick or sight glass)			
Hydraulic fluid level correct (check dip stick or sight glass)			
Hydraulic system exhibits no apparent weeping or leaks			
Air system exhibits no audible leaks			
Tire pressure acceptable and tire not damaged			
Telescoping boom exhibits no damage to structure, wear pads, boom stops, or cylinder			
Wire rope free of dirt, excess lube, kinks, and wires and spooled correctly			
Reeving correct			
Wedge sockets and wire rope clips not distorted, cracked, or missing			
Block not damaged			
Ball and hook is free to swivel and rotate			
Guards are in place			
Outrigger float(s) secured with pad pin			
Cab			
Handrails in place and not damaged			
Operator's manual in vehicle			
Load chart legible and visible to operator			
Hand signal chart visible to workers			
Charged fire extinguisher in place			
Cab glass not cracked and wipers are functional			

	Pass	Fail	N/A
Gauges and Indicators			
Load moment indicator operational			
Drum rotation indicator functioning			
Boom length indicator functioning			
Boom angle indicator functioning			
Engine: hydraulic, air, electrical, oil pressure, temperature, and fuel			
Operational Inspection			
Correct counterweight for the load			
Main hoist control functioning			
Auxiliary hoist control functioning			
Anti-two block in place and functioning			
Swing brake			
Lights and horns functional			

Notes:

Date_____

Operator
Crane number

Signature	
Model	

Visual Inspection	Pass	Fail	N/A
Engine fluid level correct (check dip stick or sight glass)			
Hydraulic fluid level correct (check dip stick or sight glass)			
Hydraulic system exhibits no apparent weeping or leaks			
Air system exhibits no audible leaks			
Tire pressure acceptable and tire not damaged			
Telescoping boom exhibits no damage to structure, wear pads, boom stops, or cylinder			
Wire rope free of dirt, excess lube, kinks, and wires and spooled correctly			
Reeving correct			
Wedge sockets and wire rope clips not distorted, cracked, or missing			
Block not damaged			
Ball and hook is free to swivel and rotate			
Guards are in place			
Outrigger float(s) secured with pad pin			
Cab			
Handrails in place and not damaged			
Operator's manual in vehicle			
Load chart legible and visible to operator			
Hand signal chart visible to workers			
Charged fire extinguisher in place			
Cab glass not cracked and wipers are functional			

	Pass	Fail	N/A
Gauges and Indicators			
Load moment indicator operational			
Drum rotation indicator functioning			
Boom length indicator functioning			
Boom angle indicator functioning			
Engine: hydraulic, air, electrical, oil pressure, temperature, and fuel			
Operational Inspection			
Correct counterweight for the load			
Main hoist control functioning			
Auxiliary hoist control functioning			
Anti-two block in place and functioning			
Swing brake			
Lights and horns functional			

Notes:

Date_____

Operator		Signature	
Crane number		Model	

Visual Inspection	Pass	Fail	N/A
Engine fluid level correct (check dip stick or sight glass)			
Hydraulic fluid level correct (check dip stick or sight glass)			
Hydraulic system exhibits no apparent weeping or leaks			
Air system exhibits no audible leaks			
Tire pressure acceptable and tire not damaged			
Telescoping boom exhibits no damage to structure, wear pads, boom stops, or cylinder			
Wire rope free of dirt, excess lube, kinks, and wires and spooled correctly			
Reeving correct			
Wedge sockets and wire rope clips not distorted, cracked, or missing			
Block not damaged			
Ball and hook is free to swivel and rotate			
Guards are in place			
Outrigger float(s) secured with pad pin			
Cab			
Handrails in place and not damaged			
Operator's manual in vehicle			
Load chart legible and visible to operator			
Hand signal chart visible to workers			
Charged fire extinguisher in place			
Cab glass not cracked and wipers are functional			

	Pass	Fail	N/A
Gauges and Indicators			
Load moment indicator operational			
Drum rotation indicator functioning			
Boom length indicator functioning			
Boom angle indicator functioning			
Engine: hydraulic, air, electrical, oil pressure, temperature, and fuel			
Operational Inspection			
Correct counterweight for the load			
Main hoist control functioning			
Auxiliary hoist control functioning			
Anti-two block in place and functioning			
Swing brake			
Lights and horns functional			

Notes:

Date_____

Operator
Crane number

Signature	
Model	

Visual Inspection	Pass	Fail	N/A
Engine fluid level correct (check dip stick or sight glass)			
Hydraulic fluid level correct (check dip stick or sight glass)			
Hydraulic system exhibits no apparent weeping or leaks			
Air system exhibits no audible leaks			
Tire pressure acceptable and tire not damaged			
Telescoping boom exhibits no damage to structure, wear pads, boom stops, or cylinder			
Wire rope free of dirt, excess lube, kinks, and wires and spooled correctly			
Reeving correct			
Wedge sockets and wire rope clips not distorted, cracked, or missing			
Block not damaged			
Ball and hook is free to swivel and rotate			
Guards are in place			
Outrigger float(s) secured with pad pin			
Cab			
Handrails in place and not damaged			
Operator's manual in vehicle			
Load chart legible and visible to operator			
Hand signal chart visible to workers			
Charged fire extinguisher in place			
Cab glass not cracked and wipers are functional			

	Pass	Fail	N/A
Gauges and Indicators			
Load moment indicator operational			
Drum rotation indicator functioning			
Boom length indicator functioning			
Boom angle indicator functioning			
Engine: hydraulic, air, electrical, oil pressure, temperature, and fuel			
Operational Inspection			
Correct counterweight for the load			
Main hoist control functioning			
Auxiliary hoist control functioning			
Anti-two block in place and functioning			
Swing brake			
Lights and horns functional			

Notes:

Date_____

Operator		Signature	
Crane number		Model	

Visual Inspection	Pass	Fail	N/A
Engine fluid level correct (check dip stick or sight glass)			
Hydraulic fluid level correct (check dip stick or sight glass)			
Hydraulic system exhibits no apparent weeping or leaks			
Air system exhibits no audible leaks			
Tire pressure acceptable and tire not damaged			
Telescoping boom exhibits no damage to structure, wear pads, boom stops, or cylinder			
Wire rope free of dirt, excess lube, kinks, and wires and spooled correctly			
Reeving correct			
Wedge sockets and wire rope clips not distorted, cracked, or missing			
Block not damaged			
Ball and hook is free to swivel and rotate			
Guards are in place			
Outrigger float(s) secured with pad pin			
Cab			
Handrails in place and not damaged			
Operator's manual in vehicle			
Load chart legible and visible to operator			
Hand signal chart visible to workers			
Charged fire extinguisher in place			
Cab glass not cracked and wipers are functional			

	Pass	Fail	N/A
Gauges and Indicators			
Load moment indicator operational			
Drum rotation indicator functioning			
Boom length indicator functioning			
Boom angle indicator functioning			
Engine: hydraulic, air, electrical, oil pressure, temperature, and fuel			
Operational Inspection			
Correct counterweight for the load			
Main hoist control functioning			
Auxiliary hoist control functioning			
Anti-two block in place and functioning			
Swing brake			
Lights and horns functional			

Notes:

Date_____

Operator
Crane number

Signature	
Model	

Visual Inspection	Pass	Fail	N/A
Engine fluid level correct (check dip stick or sight glass)			
Hydraulic fluid level correct (check dip stick or sight glass)			
Hydraulic system exhibits no apparent weeping or leaks			
Air system exhibits no audible leaks			
Tire pressure acceptable and tire not damaged			
Telescoping boom exhibits no damage to structure, wear pads, boom stops, or cylinder			
Wire rope free of dirt, excess lube, kinks, and wires and spooled correctly			
Reeving correct			
Wedge sockets and wire rope clips not distorted, cracked, or missing			
Block not damaged			
Ball and hook is free to swivel and rotate			
Guards are in place			
Outrigger float(s) secured with pad pin			
Cab			
Handrails in place and not damaged			
Operator's manual in vehicle			
Load chart legible and visible to operator			
Hand signal chart visible to workers			
Charged fire extinguisher in place			
Cab glass not cracked and wipers are functional			

	Pass	Fail	N/A
Gauges and Indicators			
Load moment indicator operational			
Drum rotation indicator functioning			
Boom length indicator functioning			
Boom angle indicator functioning			
Engine: hydraulic, air, electrical, oil pressure, temperature, and fuel			
Operational Inspection			
Correct counterweight for the load			
Main hoist control functioning			
Auxiliary hoist control functioning			
Anti-two block in place and functioning			
Swing brake			
Lights and horns functional			

Notes:

Date_____

Operator		Signature	
Crane number		Model	

Visual Inspection	Pass	Fail	N/A
Engine fluid level correct (check dip stick or sight glass)			
Hydraulic fluid level correct (check dip stick or sight glass)			
Hydraulic system exhibits no apparent weeping or leaks			
Air system exhibits no audible leaks			
Tire pressure acceptable and tire not damaged			
Telescoping boom exhibits no damage to structure, wear pads, boom stops, or cylinder			
Wire rope free of dirt, excess lube, kinks, and wires and spooled correctly			
Reeving correct			
Wedge sockets and wire rope clips not distorted, cracked, or missing			
Block not damaged			
Ball and hook is free to swivel and rotate			
Guards are in place			
Outrigger float(s) secured with pad pin			
Cab			
Handrails in place and not damaged			
Operator's manual in vehicle			
Load chart legible and visible to operator			
Hand signal chart visible to workers			
Charged fire extinguisher in place			
Cab glass not cracked and wipers are functional			

	Pass	Fail	N/A
Gauges and Indicators			
Load moment indicator operational			
Drum rotation indicator functioning			
Boom length indicator functioning			
Boom angle indicator functioning			
Engine: hydraulic, air, electrical, oil pressure, temperature, and fuel			
Operational Inspection			
Correct counterweight for the load			
Main hoist control functioning			
Auxiliary hoist control functioning			
Anti-two block in place and functioning			
Swing brake			
Lights and horns functional			

Notes:

Date_____

Operator		Signature	
Crane number		Model	

Visual Inspection	Pass	Fail	N/A
Engine fluid level correct (check dip stick or sight glass)			
Hydraulic fluid level correct (check dip stick or sight glass)			
Hydraulic system exhibits no apparent weeping or leaks			
Air system exhibits no audible leaks			
Tire pressure acceptable and tire not damaged			
Telescoping boom exhibits no damage to structure, wear pads, boom stops, or cylinder			
Wire rope free of dirt, excess lube, kinks, and wires and spooled correctly			
Reeving correct			
Wedge sockets and wire rope clips not distorted, cracked, or missing			
Block not damaged			
Ball and hook is free to swivel and rotate			
Guards are in place			
Outrigger float(s) secured with pad pin			
Cab			
Handrails in place and not damaged			
Operator's manual in vehicle			
Load chart legible and visible to operator			
Hand signal chart visible to workers			
Charged fire extinguisher in place			
Cab glass not cracked and wipers are functional			

Gauges and Indicators	Pass	Fail	N/A
Load moment indicator operational			
Drum rotation indicator functioning			
Boom length indicator functioning			
Boom angle indicator functioning			
Engine: hydraulic, air, electrical, oil pressure, temperature, and fuel			
Operational Inspection			
Correct counterweight for the load			
Main hoist control functioning			
Auxiliary hoist control functioning			
Anti-two block in place and functioning			
Swing brake			
Lights and horns functional			

Notes:

Date_____

Operator		Signature	
Crane number		Model	

Visual Inspection	Pass	Fail	N/A		Pass	Fail	N/A
Engine fluid level correct (check dip stick or sight glass)				**Gauges and Indicators**			
Hydraulic fluid level correct (check dip stick or sight glass)				Load moment indicator operational			
Hydraulic system exhibits no apparent weeping or leaks				Drum rotation indicator functioning			
Air system exhibits no audible leaks				Boom length indicator functioning			
Tire pressure acceptable and tire not damaged				Boom angle indicator functioning			
Telescoping boom exhibits no damage to structure, wear pads, boom stops, or cylinder				Engine: hydraulic, air, electrical, oil pressure, temperature, and fuel			
Wire rope free of dirt, excess lube, kinks, and wires and spooled correctly				**Operational Inspection**			
Reeving correct				Correct counterweight for the load			
Wedge sockets and wire rope clips not distorted, cracked, or missing				Main hoist control functioning			
Block not damaged				Auxiliary hoist control functioning			
Ball and hook is free to swivel and rotate				Anti-two block in place and functioning			
Guards are in place				Swing brake			
Outrigger float(s) secured with pad pin				Lights and horns functional			
Cab							
Handrails in place and not damaged							
Operator's manual in vehicle							
Load chart legible and visible to operator							
Hand signal chart visible to workers							
Charged fire extinguisher in place							
Cab glass not cracked and wipers are functional							

Notes:

Date_____

Operator
Crane number

Signature	
Model	

Visual Inspection	Pass	Fail	N/A
Engine fluid level correct (check dip stick or sight glass)			
Hydraulic fluid level correct (check dip stick or sight glass)			
Hydraulic system exhibits no apparent weeping or leaks			
Air system exhibits no audible leaks			
Tire pressure acceptable and tire not damaged			
Telescoping boom exhibits no damage to structure, wear pads, boom stops, or cylinder			
Wire rope free of dirt, excess lube, kinks, and wires and spooled correctly			
Reeving correct			
Wedge sockets and wire rope clips not distorted, cracked, or missing			
Block not damaged			
Ball and hook is free to swivel and rotate			
Guards are in place			
Outrigger float(s) secured with pad pin			
Cab			
Handrails in place and not damaged			
Operator's manual in vehicle			
Load chart legible and visible to operator			
Hand signal chart visible to workers			
Charged fire extinguisher in place			
Cab glass not cracked and wipers are functional			

	Pass	Fail	N/A
Gauges and Indicators			
Load moment indicator operational			
Drum rotation indicator functioning			
Boom length indicator functioning			
Boom angle indicator functioning			
Engine: hydraulic, air, electrical, oil pressure, temperature, and fuel			
Operational Inspection			
Correct counterweight for the load			
Main hoist control functioning			
Auxiliary hoist control functioning			
Anti-two block in place and functioning			
Swing brake			
Lights and horns functional			

Notes:

Date_____

Operator		Signature	
Crane number		Model	

Visual Inspection	Pass	Fail	N/A		Pass	Fail	N/A
Engine fluid level correct (check dip stick or sight glass)				**Gauges and Indicators**			
Hydraulic fluid level correct (check dip stick or sight glass)				Load moment indicator operational			
Hydraulic system exhibits no apparent weeping or leaks				Drum rotation indicator functioning			
Air system exhibits no audible leaks				Boom length indicator functioning			
Tire pressure acceptable and tire not damaged				Boom angle indicator functioning			
Telescoping boom exhibits no damage to structure, wear pads, boom stops, or cylinder				Engine: hydraulic, air, electrical, oil pressure, temperature, and fuel			
Wire rope free of dirt, excess lube, kinks, and wires and spooled correctly				**Operational Inspection**			
Reeving correct				Correct counterweight for the load			
Wedge sockets and wire rope clips not distorted, cracked, or missing				Main hoist control functioning			
Block not damaged				Auxiliary hoist control functioning			
Ball and hook is free to swivel and rotate				Anti-two block in place and functioning			
Guards are in place				Swing brake			
Outrigger float(s) secured with pad pin				Lights and horns functional			
Cab							
Handrails in place and not damaged							
Operator's manual in vehicle							
Load chart legible and visible to operator							
Hand signal chart visible to workers							
Charged fire extinguisher in place							
Cab glass not cracked and wipers are functional							

Notes:

Date_____

Operator		Signature	
Crane number		Model	

Visual Inspection	Pass	Fail	N/A
Engine fluid level correct (check dip stick or sight glass)			
Hydraulic fluid level correct (check dip stick or sight glass)			
Hydraulic system exhibits no apparent weeping or leaks			
Air system exhibits no audible leaks			
Tire pressure acceptable and tire not damaged			
Telescoping boom exhibits no damage to structure, wear pads, boom stops, or cylinder			
Wire rope free of dirt, excess lube, kinks, and wires and spooled correctly			
Reeving correct			
Wedge sockets and wire rope clips not distorted, cracked, or missing			
Block not damaged			
Ball and hook is free to swivel and rotate			
Guards are in place			
Outrigger float(s) secured with pad pin			
Cab			
Handrails in place and not damaged			
Operator's manual in vehicle			
Load chart legible and visible to operator			
Hand signal chart visible to workers			
Charged fire extinguisher in place			
Cab glass not cracked and wipers are functional			

	Pass	Fail	N/A
Gauges and Indicators			
Load moment indicator operational			
Drum rotation indicator functioning			
Boom length indicator functioning			
Boom angle indicator functioning			
Engine: hydraulic, air, electrical, oil pressure, temperature, and fuel			
Operational Inspection			
Correct counterweight for the load			
Main hoist control functioning			
Auxiliary hoist control functioning			
Anti-two block in place and functioning			
Swing brake			
Lights and horns functional			

Notes:

Date _____

Operator		Signature	
Crane number		Model	

Visual Inspection	Pass	Fail	N/A		Pass	Fail	N/A
Engine fluid level correct (check dip stick or sight glass)				**Gauges and Indicators**			
Hydraulic fluid level correct (check dip stick or sight glass)				Load moment indicator operational			
Hydraulic system exhibits no apparent weeping or leaks				Drum rotation indicator functioning			
Air system exhibits no audible leaks				Boom length indicator functioning			
Tire pressure acceptable and tire not damaged				Boom angle indicator functioning			
Telescoping boom exhibits no damage to structure, wear pads, boom stops, or cylinder				Engine: hydraulic, air, electrical, oil pressure, temperature, and fuel			
Wire rope free of dirt, excess lube, kinks, and wires and spooled correctly				**Operational Inspection**			
Reeving correct				Correct counterweight for the load			
Wedge sockets and wire rope clips not distorted, cracked, or missing				Main hoist control functioning			
Block not damaged				Auxiliary hoist control functioning			
Ball and hook is free to swivel and rotate				Anti-two block in place and functioning			
Guards are in place				Swing brake			
Outrigger float(s) secured with pad pin				Lights and horns functional			
Cab							
Handrails in place and not damaged							
Operator's manual in vehicle							
Load chart legible and visible to operator							
Hand signal chart visible to workers							
Charged fire extinguisher in place							
Cab glass not cracked and wipers are functional							

Notes:

Date_____

Operator
Crane number

Signature	
Model	

Visual Inspection	Pass	Fail	N/A
Engine fluid level correct (check dip stick or sight glass)			
Hydraulic fluid level correct (check dip stick or sight glass)			
Hydraulic system exhibits no apparent weeping or leaks			
Air system exhibits no audible leaks			
Tire pressure acceptable and tire not damaged			
Telescoping boom exhibits no damage to structure, wear pads, boom stops, or cylinder			
Wire rope free of dirt, excess lube, kinks, and wires and spooled correctly			
Reeving correct			
Wedge sockets and wire rope clips not distorted, cracked, or missing			
Block not damaged			
Ball and hook is free to swivel and rotate			
Guards are in place			
Outrigger float(s) secured with pad pin			
Cab			
Handrails in place and not damaged			
Operator's manual in vehicle			
Load chart legible and visible to operator			
Hand signal chart visible to workers			
Charged fire extinguisher in place			
Cab glass not cracked and wipers are functional			

	Pass	Fail	N/A
Gauges and Indicators			
Load moment indicator operational			
Drum rotation indicator functioning			
Boom length indicator functioning			
Boom angle indicator functioning			
Engine: hydraulic, air, electrical, oil pressure, temperature, and fuel			
Operational Inspection			
Correct counterweight for the load			
Main hoist control functioning			
Auxiliary hoist control functioning			
Anti-two block in place and functioning			
Swing brake			
Lights and horns functional			

Notes:

Date_____

Operator		Signature	
Crane number		Model	

Visual Inspection	Pass	Fail	N/A
Engine fluid level correct (check dip stick or sight glass)			
Hydraulic fluid level correct (check dip stick or sight glass)			
Hydraulic system exhibits no apparent weeping or leaks			
Air system exhibits no audible leaks			
Tire pressure acceptable and tire not damaged			
Telescoping boom exhibits no damage to structure, wear pads, boom stops, or cylinder			
Wire rope free of dirt, excess lube, kinks, and wires and spooled correctly			
Reeving correct			
Wedge sockets and wire rope clips not distorted, cracked, or missing			
Block not damaged			
Ball and hook is free to swivel and rotate			
Guards are in place			
Outrigger float(s) secured with pad pin			
Cab			
Handrails in place and not damaged			
Operator's manual in vehicle			
Load chart legible and visible to operator			
Hand signal chart visible to workers			
Charged fire extinguisher in place			
Cab glass not cracked and wipers are functional			

	Pass	Fail	N/A
Gauges and Indicators			
Load moment indicator operational			
Drum rotation indicator functioning			
Boom length indicator functioning			
Boom angle indicator functioning			
Engine: hydraulic, air, electrical, oil pressure, temperature, and fuel			
Operational Inspection			
Correct counterweight for the load			
Main hoist control functioning			
Auxiliary hoist control functioning			
Anti-two block in place and functioning			
Swing brake			
Lights and horns functional			

Notes:

Date_____

Operator
Crane number

Signature	
Model	

Visual Inspection	Pass	Fail	N/A
Engine fluid level correct (check dip stick or sight glass)			
Hydraulic fluid level correct (check dip stick or sight glass)			
Hydraulic system exhibits no apparent weeping or leaks			
Air system exhibits no audible leaks			
Tire pressure acceptable and tire not damaged			
Telescoping boom exhibits no damage to structure, wear pads, boom stops, or cylinder			
Wire rope free of dirt, excess lube, kinks, and wires and spooled correctly			
Reeving correct			
Wedge sockets and wire rope clips not distorted, cracked, or missing			
Block not damaged			
Ball and hook is free to swivel and rotate			
Guards are in place			
Outrigger float(s) secured with pad pin			
Cab			
Handrails in place and not damaged			
Operator's manual in vehicle			
Load chart legible and visible to operator			
Hand signal chart visible to workers			
Charged fire extinguisher in place			
Cab glass not cracked and wipers are functional			

	Pass	Fail	N/A
Gauges and Indicators			
Load moment indicator operational			
Drum rotation indicator functioning			
Boom length indicator functioning			
Boom angle indicator functioning			
Engine: hydraulic, air, electrical, oil pressure, temperature, and fuel			
Operational Inspection			
Correct counterweight for the load			
Main hoist control functioning			
Auxiliary hoist control functioning			
Anti-two block in place and functioning			
Swing brake			
Lights and horns functional			

Notes:

Date_____

Operator
Crane number

Signature	
Model	

Visual Inspection	Pass	Fail	N/A
Engine fluid level correct (check dip stick or sight glass)			
Hydraulic fluid level correct (check dip stick or sight glass)			
Hydraulic system exhibits no apparent weeping or leaks			
Air system exhibits no audible leaks			
Tire pressure acceptable and tire not damaged			
Telescoping boom exhibits no damage to structure, wear pads, boom stops, or cylinder			
Wire rope free of dirt, excess lube, kinks, and wires and spooled correctly			
Reeving correct			
Wedge sockets and wire rope clips not distorted, cracked, or missing			
Block not damaged			
Ball and hook is free to swivel and rotate			
Guards are in place			
Outrigger float(s) secured with pad pin			
Cab			
Handrails in place and not damaged			
Operator's manual in vehicle			
Load chart legible and visible to operator			
Hand signal chart visible to workers			
Charged fire extinguisher in place			
Cab glass not cracked and wipers are functional			

	Pass	Fail	N/A
Gauges and Indicators			
Load moment indicator operational			
Drum rotation indicator functioning			
Boom length indicator functioning			
Boom angle indicator functioning			
Engine: hydraulic, air, electrical, oil pressure, temperature, and fuel			
Operational Inspection			
Correct counterweight for the load			
Main hoist control functioning			
Auxiliary hoist control functioning			
Anti-two block in place and functioning			
Swing brake			
Lights and horns functional			

Notes:

Date_____

Operator
Crane number

Signature	
Model	

Visual Inspection	Pass	Fail	N/A
Engine fluid level correct (check dip stick or sight glass)			
Hydraulic fluid level correct (check dip stick or sight glass)			
Hydraulic system exhibits no apparent weeping or leaks			
Air system exhibits no audible leaks			
Tire pressure acceptable and tire not damaged			
Telescoping boom exhibits no damage to structure, wear pads, boom stops, or cylinder			
Wire rope free of dirt, excess lube, kinks, and wires and spooled correctly			
Reeving correct			
Wedge sockets and wire rope clips not distorted, cracked, or missing			
Block not damaged			
Ball and hook is free to swivel and rotate			
Guards are in place			
Outrigger float(s) secured with pad pin			
Cab			
Handrails in place and not damaged			
Operator's manual in vehicle			
Load chart legible and visible to operator			
Hand signal chart visible to workers			
Charged fire extinguisher in place			
Cab glass not cracked and wipers are functional			

	Pass	Fail	N/A
Gauges and Indicators			
Load moment indicator operational			
Drum rotation indicator functioning			
Boom length indicator functioning			
Boom angle indicator functioning			
Engine: hydraulic, air, electrical, oil pressure, temperature, and fuel			
Operational Inspection			
Correct counterweight for the load			
Main hoist control functioning			
Auxiliary hoist control functioning			
Anti-two block in place and functioning			
Swing brake			
Lights and horns functional			

Notes:

Date_____

Operator		Signature	
Crane number		Model	

Visual Inspection	Pass	Fail	N/A
Engine fluid level correct (check dip stick or sight glass)			
Hydraulic fluid level correct (check dip stick or sight glass)			
Hydraulic system exhibits no apparent weeping or leaks			
Air system exhibits no audible leaks			
Tire pressure acceptable and tire not damaged			
Telescoping boom exhibits no damage to structure, wear pads, boom stops, or cylinder			
Wire rope free of dirt, excess lube, kinks, and wires and spooled correctly			
Reeving correct			
Wedge sockets and wire rope clips not distorted, cracked, or missing			
Block not damaged			
Ball and hook is free to swivel and rotate			
Guards are in place			
Outrigger float(s) secured with pad pin			
Cab			
Handrails in place and not damaged			
Operator's manual in vehicle			
Load chart legible and visible to operator			
Hand signal chart visible to workers			
Charged fire extinguisher in place			
Cab glass not cracked and wipers are functional			

	Pass	Fail	N/A
Gauges and Indicators			
Load moment indicator operational			
Drum rotation indicator functioning			
Boom length indicator functioning			
Boom angle indicator functioning			
Engine: hydraulic, air, electrical, oil pressure, temperature, and fuel			
Operational Inspection			
Correct counterweight for the load			
Main hoist control functioning			
Auxiliary hoist control functioning			
Anti-two block in place and functioning			
Swing brake			
Lights and horns functional			

Notes:

Date_____

Operator	
Crane number	

Signature		
Model		

Visual Inspection	Pass	Fail	N/A
Engine fluid level correct (check dip stick or sight glass)			
Hydraulic fluid level correct (check dip stick or sight glass)			
Hydraulic system exhibits no apparent weeping or leaks			
Air system exhibits no audible leaks			
Tire pressure acceptable and tire not damaged			
Telescoping boom exhibits no damage to structure, wear pads, boom stops, or cylinder			
Wire rope free of dirt, excess lube, kinks, and wires and spooled correctly			
Reeving correct			
Wedge sockets and wire rope clips not distorted, cracked, or missing			
Block not damaged			
Ball and hook is free to swivel and rotate			
Guards are in place			
Outrigger float(s) secured with pad pin			
Cab			
Handrails in place and not damaged			
Operator's manual in vehicle			
Load chart legible and visible to operator			
Hand signal chart visible to workers			
Charged fire extinguisher in place			
Cab glass not cracked and wipers are functional			

	Pass	Fail	N/A
Gauges and Indicators			
Load moment indicator operational			
Drum rotation indicator functioning			
Boom length indicator functioning			
Boom angle indicator functioning			
Engine: hydraulic, air, electrical, oil pressure, temperature, and fuel			
Operational Inspection			
Correct counterweight for the load			
Main hoist control functioning			
Auxiliary hoist control functioning			
Anti-two block in place and functioning			
Swing brake			
Lights and horns functional			

Notes:

Date_____

Operator		Signature	
Crane number		Model	

Visual Inspection	Pass	Fail	N/A
Engine fluid level correct (check dip stick or sight glass)			
Hydraulic fluid level correct (check dip stick or sight glass)			
Hydraulic system exhibits no apparent weeping or leaks			
Air system exhibits no audible leaks			
Tire pressure acceptable and tire not damaged			
Telescoping boom exhibits no damage to structure, wear pads, boom stops, or cylinder			
Wire rope free of dirt, excess lube, kinks, and wires and spooled correctly			
Reeving correct			
Wedge sockets and wire rope clips not distorted, cracked, or missing			
Block not damaged			
Ball and hook is free to swivel and rotate			
Guards are in place			
Outrigger float(s) secured with pad pin			
Cab			
Handrails in place and not damaged			
Operator's manual in vehicle			
Load chart legible and visible to operator			
Hand signal chart visible to workers			
Charged fire extinguisher in place			
Cab glass not cracked and wipers are functional			

	Pass	Fail	N/A
Gauges and Indicators			
Load moment indicator operational			
Drum rotation indicator functioning			
Boom length indicator functioning			
Boom angle indicator functioning			
Engine: hydraulic, air, electrical, oil pressure, temperature, and fuel			
Operational Inspection			
Correct counterweight for the load			
Main hoist control functioning			
Auxiliary hoist control functioning			
Anti-two block in place and functioning			
Swing brake			
Lights and horns functional			

Notes:

Date _____

Operator
Crane number

Signature	
Model	

Visual Inspection	Pass	Fail	N/A
Engine fluid level correct (check dip stick or sight glass)			
Hydraulic fluid level correct (check dip stick or sight glass)			
Hydraulic system exhibits no apparent weeping or leaks			
Air system exhibits no audible leaks			
Tire pressure acceptable and tire not damaged			
Telescoping boom exhibits no damage to structure, wear pads, boom stops, or cylinder			
Wire rope free of dirt, excess lube, kinks, and wires and spooled correctly			
Reeving correct			
Wedge sockets and wire rope clips not distorted, cracked, or missing			
Block not damaged			
Ball and hook is free to swivel and rotate			
Guards are in place			
Outrigger float(s) secured with pad pin			
Cab			
Handrails in place and not damaged			
Operator's manual in vehicle			
Load chart legible and visible to operator			
Hand signal chart visible to workers			
Charged fire extinguisher in place			
Cab glass not cracked and wipers are functional			

	Pass	Fail	N/A
Gauges and Indicators			
Load moment indicator operational			
Drum rotation indicator functioning			
Boom length indicator functioning			
Boom angle indicator functioning			
Engine: hydraulic, air, electrical, oil pressure, temperature, and fuel			
Operational Inspection			
Correct counterweight for the load			
Main hoist control functioning			
Auxiliary hoist control functioning			
Anti-two block in place and functioning			
Swing brake			
Lights and horns functional			

Notes:

Date_____

Operator	
Crane number	

Signature		
Model		

Visual Inspection	Pass	Fail	N/A
Engine fluid level correct (check dip stick or sight glass)			
Hydraulic fluid level correct (check dip stick or sight glass)			
Hydraulic system exhibits no apparent weeping or leaks			
Air system exhibits no audible leaks			
Tire pressure acceptable and tire not damaged			
Telescoping boom exhibits no damage to structure, wear pads, boom stops, or cylinder			
Wire rope free of dirt, excess lube, kinks, and wires and spooled correctly			
Reeving correct			
Wedge sockets and wire rope clips not distorted, cracked, or missing			
Block not damaged			
Ball and hook is free to swivel and rotate			
Guards are in place			
Outrigger float(s) secured with pad pin			
Cab			
Handrails in place and not damaged			
Operator's manual in vehicle			
Load chart legible and visible to operator			
Hand signal chart visible to workers			
Charged fire extinguisher in place			
Cab glass not cracked and wipers are functional			

	Pass	Fail	N/A
Gauges and Indicators			
Load moment indicator operational			
Drum rotation indicator functioning			
Boom length indicator functioning			
Boom angle indicator functioning			
Engine: hydraulic, air, electrical, oil pressure, temperature, and fuel			
Operational Inspection			
Correct counterweight for the load			
Main hoist control functioning			
Auxiliary hoist control functioning			
Anti-two block in place and functioning			
Swing brake			
Lights and horns functional			

Notes:

Date_____

Operator
Crane number

Signature	
Model	

Visual Inspection	Pass	Fail	N/A
Engine fluid level correct (check dip stick or sight glass)			
Hydraulic fluid level correct (check dip stick or sight glass)			
Hydraulic system exhibits no apparent weeping or leaks			
Air system exhibits no audible leaks			
Tire pressure acceptable and tire not damaged			
Telescoping boom exhibits no damage to structure, wear pads, boom stops, or cylinder			
Wire rope free of dirt, excess lube, kinks, and wires and spooled correctly			
Reeving correct			
Wedge sockets and wire rope clips not distorted, cracked, or missing			
Block not damaged			
Ball and hook is free to swivel and rotate			
Guards are in place			
Outrigger float(s) secured with pad pin			
Cab			
Handrails in place and not damaged			
Operator's manual in vehicle			
Load chart legible and visible to operator			
Hand signal chart visible to workers			
Charged fire extinguisher in place			
Cab glass not cracked and wipers are functional			

	Pass	Fail	N/A
Gauges and Indicators			
Load moment indicator operational			
Drum rotation indicator functioning			
Boom length indicator functioning			
Boom angle indicator functioning			
Engine: hydraulic, air, electrical, oil pressure, temperature, and fuel			
Operational Inspection			
Correct counterweight for the load			
Main hoist control functioning			
Auxiliary hoist control functioning			
Anti-two block in place and functioning			
Swing brake			
Lights and horns functional			

Notes:

Date_____

Operator
Crane number

Signature	
Model	

Visual Inspection	Pass	Fail	N/A
Engine fluid level correct (check dip stick or sight glass)			
Hydraulic fluid level correct (check dip stick or sight glass)			
Hydraulic system exhibits no apparent weeping or leaks			
Air system exhibits no audible leaks			
Tire pressure acceptable and tire not damaged			
Telescoping boom exhibits no damage to structure, wear pads, boom stops, or cylinder			
Wire rope free of dirt, excess lube, kinks, and wires and spooled correctly			
Reeving correct			
Wedge sockets and wire rope clips not distorted, cracked, or missing			
Block not damaged			
Ball and hook is free to swivel and rotate			
Guards are in place			
Outrigger float(s) secured with pad pin			
Cab			
Handrails in place and not damaged			
Operator's manual in vehicle			
Load chart legible and visible to operator			
Hand signal chart visible to workers			
Charged fire extinguisher in place			
Cab glass not cracked and wipers are functional			

	Pass	Fail	N/A
Gauges and Indicators			
Load moment indicator operational			
Drum rotation indicator functioning			
Boom length indicator functioning			
Boom angle indicator functioning			
Engine: hydraulic, air, electrical, oil pressure, temperature, and fuel			
Operational Inspection			
Correct counterweight for the load			
Main hoist control functioning			
Auxiliary hoist control functioning			
Anti-two block in place and functioning			
Swing brake			
Lights and horns functional			

Notes:

Date _____

Operator
Crane number

Signature	
Model	

Visual Inspection	Pass	Fail	N/A
Engine fluid level correct (check dip stick or sight glass)			
Hydraulic fluid level correct (check dip stick or sight glass)			
Hydraulic system exhibits no apparent weeping or leaks			
Air system exhibits no audible leaks			
Tire pressure acceptable and tire not damaged			
Telescoping boom exhibits no damage to structure, wear pads, boom stops, or cylinder			
Wire rope free of dirt, excess lube, kinks, and wires and spooled correctly			
Reeving correct			
Wedge sockets and wire rope clips not distorted, cracked, or missing			
Block not damaged			
Ball and hook is free to swivel and rotate			
Guards are in place			
Outrigger float(s) secured with pad pin			
Cab			
Handrails in place and not damaged			
Operator's manual in vehicle			
Load chart legible and visible to operator			
Hand signal chart visible to workers			
Charged fire extinguisher in place			
Cab glass not cracked and wipers are functional			

	Pass	Fail	N/A
Gauges and Indicators			
Load moment indicator operational			
Drum rotation indicator functioning			
Boom length indicator functioning			
Boom angle indicator functioning			
Engine: hydraulic, air, electrical, oil pressure, temperature, and fuel			
Operational Inspection			
Correct counterweight for the load			
Main hoist control functioning			
Auxiliary hoist control functioning			
Anti-two block in place and functioning			
Swing brake			
Lights and horns functional			

Notes:

Date_____

Operator		Signature	
Crane number		Model	

Visual Inspection	Pass	Fail	N/A
Engine fluid level correct (check dip stick or sight glass)			
Hydraulic fluid level correct (check dip stick or sight glass)			
Hydraulic system exhibits no apparent weeping or leaks			
Air system exhibits no audible leaks			
Tire pressure acceptable and tire not damaged			
Telescoping boom exhibits no damage to structure, wear pads, boom stops, or cylinder			
Wire rope free of dirt, excess lube, kinks, and wires and spooled correctly			
Reeving correct			
Wedge sockets and wire rope clips not distorted, cracked, or missing			
Block not damaged			
Ball and hook is free to swivel and rotate			
Guards are in place			
Outrigger float(s) secured with pad pin			
Cab			
Handrails in place and not damaged			
Operator's manual in vehicle			
Load chart legible and visible to operator			
Hand signal chart visible to workers			
Charged fire extinguisher in place			
Cab glass not cracked and wipers are functional			

	Pass	Fail	N/A
Gauges and Indicators			
Load moment indicator operational			
Drum rotation indicator functioning			
Boom length indicator functioning			
Boom angle indicator functioning			
Engine: hydraulic, air, electrical, oil pressure, temperature, and fuel			
Operational Inspection			
Correct counterweight for the load			
Main hoist control functioning			
Auxiliary hoist control functioning			
Anti-two block in place and functioning			
Swing brake			
Lights and horns functional			

Notes:

Date_____

Operator	
Crane number	

Signature	
Model	

Visual Inspection	Pass	Fail	N/A
Engine fluid level correct (check dip stick or sight glass)			
Hydraulic fluid level correct (check dip stick or sight glass)			
Hydraulic system exhibits no apparent weeping or leaks			
Air system exhibits no audible leaks			
Tire pressure acceptable and tire not damaged			
Telescoping boom exhibits no damage to structure, wear pads, boom stops, or cylinder			
Wire rope free of dirt, excess lube, kinks, and wires and spooled correctly			
Reeving correct			
Wedge sockets and wire rope clips not distorted, cracked, or missing			
Block not damaged			
Ball and hook is free to swivel and rotate			
Guards are in place			
Outrigger float(s) secured with pad pin			
Cab			
Handrails in place and not damaged			
Operator's manual in vehicle			
Load chart legible and visible to operator			
Hand signal chart visible to workers			
Charged fire extinguisher in place			
Cab glass not cracked and wipers are functional			

	Pass	Fail	N/A
Gauges and Indicators			
Load moment indicator operational			
Drum rotation indicator functioning			
Boom length indicator functioning			
Boom angle indicator functioning			
Engine: hydraulic, air, electrical, oil pressure, temperature, and fuel			
Operational Inspection			
Correct counterweight for the load			
Main hoist control functioning			
Auxiliary hoist control functioning			
Anti-two block in place and functioning			
Swing brake			
Lights and horns functional			

Notes:

Date_____

Operator
Crane number

Signature	
Model	

Visual Inspection	Pass	Fail	N/A
Engine fluid level correct (check dip stick or sight glass)			
Hydraulic fluid level correct (check dip stick or sight glass)			
Hydraulic system exhibits no apparent weeping or leaks			
Air system exhibits no audible leaks			
Tire pressure acceptable and tire not damaged			
Telescoping boom exhibits no damage to structure, wear pads, boom stops, or cylinder			
Wire rope free of dirt, excess lube, kinks, and wires and spooled correctly			
Reeving correct			
Wedge sockets and wire rope clips not distorted, cracked, or missing			
Block not damaged			
Ball and hook is free to swivel and rotate			
Guards are in place			
Outrigger float(s) secured with pad pin			
Cab			
Handrails in place and not damaged			
Operator's manual in vehicle			
Load chart legible and visible to operator			
Hand signal chart visible to workers			
Charged fire extinguisher in place			
Cab glass not cracked and wipers are functional			

	Pass	Fail	N/A
Gauges and Indicators			
Load moment indicator operational			
Drum rotation indicator functioning			
Boom length indicator functioning			
Boom angle indicator functioning			
Engine: hydraulic, air, electrical, oil pressure, temperature, and fuel			
Operational Inspection			
Correct counterweight for the load			
Main hoist control functioning			
Auxiliary hoist control functioning			
Anti-two block in place and functioning			
Swing brake			
Lights and horns functional			

Notes:

Date_____

Operator
Crane number

Signature	
Model	

Visual Inspection	Pass	Fail	N/A
Engine fluid level correct (check dip stick or sight glass)			
Hydraulic fluid level correct (check dip stick or sight glass)			
Hydraulic system exhibits no apparent weeping or leaks			
Air system exhibits no audible leaks			
Tire pressure acceptable and tire not damaged			
Telescoping boom exhibits no damage to structure, wear pads, boom stops, or cylinder			
Wire rope free of dirt, excess lube, kinks, and wires and spooled correctly			
Reeving correct			
Wedge sockets and wire rope clips not distorted, cracked, or missing			
Block not damaged			
Ball and hook is free to swivel and rotate			
Guards are in place			
Outrigger float(s) secured with pad pin			
Cab			
Handrails in place and not damaged			
Operator's manual in vehicle			
Load chart legible and visible to operator			
Hand signal chart visible to workers			
Charged fire extinguisher in place			
Cab glass not cracked and wipers are functional			

	Pass	Fail	N/A
Gauges and Indicators			
Load moment indicator operational			
Drum rotation indicator functioning			
Boom length indicator functioning			
Boom angle indicator functioning			
Engine: hydraulic, air, electrical, oil pressure, temperature, and fuel			
Operational Inspection			
Correct counterweight for the load			
Main hoist control functioning			
Auxiliary hoist control functioning			
Anti-two block in place and functioning			
Swing brake			
Lights and horns functional			

Notes:

Date _____

Operator		Signature	
Crane number		Model	

Visual Inspection	Pass	Fail	N/A
Engine fluid level correct (check dip stick or sight glass)			
Hydraulic fluid level correct (check dip stick or sight glass)			
Hydraulic system exhibits no apparent weeping or leaks			
Air system exhibits no audible leaks			
Tire pressure acceptable and tire not damaged			
Telescoping boom exhibits no damage to structure, wear pads, boom stops, or cylinder			
Wire rope free of dirt, excess lube, kinks, and wires and spooled correctly			
Reeving correct			
Wedge sockets and wire rope clips not distorted, cracked, or missing			
Block not damaged			
Ball and hook is free to swivel and rotate			
Guards are in place			
Outrigger float(s) secured with pad pin			
Cab			
Handrails in place and not damaged			
Operator's manual in vehicle			
Load chart legible and visible to operator			
Hand signal chart visible to workers			
Charged fire extinguisher in place			
Cab glass not cracked and wipers are functional			

	Pass	Fail	N/A
Gauges and Indicators			
Load moment indicator operational			
Drum rotation indicator functioning			
Boom length indicator functioning			
Boom angle indicator functioning			
Engine: hydraulic, air, electrical, oil pressure, temperature, and fuel			
Operational Inspection			
Correct counterweight for the load			
Main hoist control functioning			
Auxiliary hoist control functioning			
Anti-two block in place and functioning			
Swing brake			
Lights and horns functional			

Notes:

Date_____

Operator
Crane number

Signature	
Model	

Visual Inspection	Pass	Fail	N/A
Engine fluid level correct (check dip stick or sight glass)			
Hydraulic fluid level correct (check dip stick or sight glass)			
Hydraulic system exhibits no apparent weeping or leaks			
Air system exhibits no audible leaks			
Tire pressure acceptable and tire not damaged			
Telescoping boom exhibits no damage to structure, wear pads, boom stops, or cylinder			
Wire rope free of dirt, excess lube, kinks, and wires and spooled correctly			
Reeving correct			
Wedge sockets and wire rope clips not distorted, cracked, or missing			
Block not damaged			
Ball and hook is free to swivel and rotate			
Guards are in place			
Outrigger float(s) secured with pad pin			
Cab			
Handrails in place and not damaged			
Operator's manual in vehicle			
Load chart legible and visible to operator			
Hand signal chart visible to workers			
Charged fire extinguisher in place			
Cab glass not cracked and wipers are functional			

	Pass	Fail	N/A
Gauges and Indicators			
Load moment indicator operational			
Drum rotation indicator functioning			
Boom length indicator functioning			
Boom angle indicator functioning			
Engine: hydraulic, air, electrical, oil pressure, temperature, and fuel			
Operational Inspection			
Correct counterweight for the load			
Main hoist control functioning			
Auxiliary hoist control functioning			
Anti-two block in place and functioning			
Swing brake			
Lights and horns functional			

Notes:

Date_____

Operator		Signature	
Crane number		Model	

Visual Inspection	Pass	Fail	N/A
Engine fluid level correct (check dip stick or sight glass)			
Hydraulic fluid level correct (check dip stick or sight glass)			
Hydraulic system exhibits no apparent weeping or leaks			
Air system exhibits no audible leaks			
Tire pressure acceptable and tire not damaged			
Telescoping boom exhibits no damage to structure, wear pads, boom stops, or cylinder			
Wire rope free of dirt, excess lube, kinks, and wires and spooled correctly			
Reeving correct			
Wedge sockets and wire rope clips not distorted, cracked, or missing			
Block not damaged			
Ball and hook is free to swivel and rotate			
Guards are in place			
Outrigger float(s) secured with pad pin			
Cab			
Handrails in place and not damaged			
Operator's manual in vehicle			
Load chart legible and visible to operator			
Hand signal chart visible to workers			
Charged fire extinguisher in place			
Cab glass not cracked and wipers are functional			

	Pass	Fail	N/A
Gauges and Indicators			
Load moment indicator operational			
Drum rotation indicator functioning			
Boom length indicator functioning			
Boom angle indicator functioning			
Engine: hydraulic, air, electrical, oil pressure, temperature, and fuel			
Operational Inspection			
Correct counterweight for the load			
Main hoist control functioning			
Auxiliary hoist control functioning			
Anti-two block in place and functioning			
Swing brake			
Lights and horns functional			

Notes:

Date _____

Operator		Signature	
Crane number		Model	

Visual Inspection	Pass	Fail	N/A
Engine fluid level correct (check dip stick or sight glass)			
Hydraulic fluid level correct (check dip stick or sight glass)			
Hydraulic system exhibits no apparent weeping or leaks			
Air system exhibits no audible leaks			
Tire pressure acceptable and tire not damaged			
Telescoping boom exhibits no damage to structure, wear pads, boom stops, or cylinder			
Wire rope free of dirt, excess lube, kinks, and wires and spooled correctly			
Reeving correct			
Wedge sockets and wire rope clips not distorted, cracked, or missing			
Block not damaged			
Ball and hook is free to swivel and rotate			
Guards are in place			
Outrigger float(s) secured with pad pin			
Cab			
Handrails in place and not damaged			
Operator's manual in vehicle			
Load chart legible and visible to operator			
Hand signal chart visible to workers			
Charged fire extinguisher in place			
Cab glass not cracked and wipers are functional			

	Pass	Fail	N/A
Gauges and Indicators			
Load moment indicator operational			
Drum rotation indicator functioning			
Boom length indicator functioning			
Boom angle indicator functioning			
Engine: hydraulic, air, electrical, oil pressure, temperature, and fuel			
Operational Inspection			
Correct counterweight for the load			
Main hoist control functioning			
Auxiliary hoist control functioning			
Anti-two block in place and functioning			
Swing brake			
Lights and horns functional			

Notes:

Date_____

Operator		Signature	
Crane number		Model	

Visual Inspection	Pass	Fail	N/A
Engine fluid level correct (check dip stick or sight glass)			
Hydraulic fluid level correct (check dip stick or sight glass)			
Hydraulic system exhibits no apparent weeping or leaks			
Air system exhibits no audible leaks			
Tire pressure acceptable and tire not damaged			
Telescoping boom exhibits no damage to structure, wear pads, boom stops, or cylinder			
Wire rope free of dirt, excess lube, kinks, and wires and spooled correctly			
Reeving correct			
Wedge sockets and wire rope clips not distorted, cracked, or missing			
Block not damaged			
Ball and hook is free to swivel and rotate			
Guards are in place			
Outrigger float(s) secured with pad pin			
Cab			
Handrails in place and not damaged			
Operator's manual in vehicle			
Load chart legible and visible to operator			
Hand signal chart visible to workers			
Charged fire extinguisher in place			
Cab glass not cracked and wipers are functional			

	Pass	Fail	N/A
Gauges and Indicators			
Load moment indicator operational			
Drum rotation indicator functioning			
Boom length indicator functioning			
Boom angle indicator functioning			
Engine: hydraulic, air, electrical, oil pressure, temperature, and fuel			
Operational Inspection			
Correct counterweight for the load			
Main hoist control functioning			
Auxiliary hoist control functioning			
Anti-two block in place and functioning			
Swing brake			
Lights and horns functional			

Notes:

Date_____

Operator
Crane number

Signature	
Model	

Visual Inspection	Pass	Fail	N/A
Engine fluid level correct (check dip stick or sight glass)			
Hydraulic fluid level correct (check dip stick or sight glass)			
Hydraulic system exhibits no apparent weeping or leaks			
Air system exhibits no audible leaks			
Tire pressure acceptable and tire not damaged			
Telescoping boom exhibits no damage to structure, wear pads, boom stops, or cylinder			
Wire rope free of dirt, excess lube, kinks, and wires and spooled correctly			
Reeving correct			
Wedge sockets and wire rope clips not distorted, cracked, or missing			
Block not damaged			
Ball and hook is free to swivel and rotate			
Guards are in place			
Outrigger float(s) secured with pad pin			
Cab			
Handrails in place and not damaged			
Operator's manual in vehicle			
Load chart legible and visible to operator			
Hand signal chart visible to workers			
Charged fire extinguisher in place			
Cab glass not cracked and wipers are functional			

	Pass	Fail	N/A
Gauges and Indicators			
Load moment indicator operational			
Drum rotation indicator functioning			
Boom length indicator functioning			
Boom angle indicator functioning			
Engine: hydraulic, air, electrical, oil pressure, temperature, and fuel			
Operational Inspection			
Correct counterweight for the load			
Main hoist control functioning			
Auxiliary hoist control functioning			
Anti-two block in place and functioning			
Swing brake			
Lights and horns functional			

Notes:

Date_____

Operator
Crane number

Signature	
Model	

Visual Inspection	Pass	Fail	N/A
Engine fluid level correct (check dip stick or sight glass)			
Hydraulic fluid level correct (check dip stick or sight glass)			
Hydraulic system exhibits no apparent weeping or leaks			
Air system exhibits no audible leaks			
Tire pressure acceptable and tire not damaged			
Telescoping boom exhibits no damage to structure, wear pads, boom stops, or cylinder			
Wire rope free of dirt, excess lube, kinks, and wires and spooled correctly			
Reeving correct			
Wedge sockets and wire rope clips not distorted, cracked, or missing			
Block not damaged			
Ball and hook is free to swivel and rotate			
Guards are in place			
Outrigger float(s) secured with pad pin			
Cab			
Handrails in place and not damaged			
Operator's manual in vehicle			
Load chart legible and visible to operator			
Hand signal chart visible to workers			
Charged fire extinguisher in place			
Cab glass not cracked and wipers are functional			

	Pass	Fail	N/A
Gauges and Indicators			
Load moment indicator operational			
Drum rotation indicator functioning			
Boom length indicator functioning			
Boom angle indicator functioning			
Engine: hydraulic, air, electrical, oil pressure, temperature, and fuel			
Operational Inspection			
Correct counterweight for the load			
Main hoist control functioning			
Auxiliary hoist control functioning			
Anti-two block in place and functioning			
Swing brake			
Lights and horns functional			

Notes:

Date_____

Operator
Crane number

Signature	
Model	

Visual Inspection	Pass	Fail	N/A
Engine fluid level correct (check dip stick or sight glass)			
Hydraulic fluid level correct (check dip stick or sight glass)			
Hydraulic system exhibits no apparent weeping or leaks			
Air system exhibits no audible leaks			
Tire pressure acceptable and tire not damaged			
Telescoping boom exhibits no damage to structure, wear pads, boom stops, or cylinder			
Wire rope free of dirt, excess lube, kinks, and wires and spooled correctly			
Reeving correct			
Wedge sockets and wire rope clips not distorted, cracked, or missing			
Block not damaged			
Ball and hook is free to swivel and rotate			
Guards are in place			
Outrigger float(s) secured with pad pin			
Cab			
Handrails in place and not damaged			
Operator's manual in vehicle			
Load chart legible and visible to operator			
Hand signal chart visible to workers			
Charged fire extinguisher in place			
Cab glass not cracked and wipers are functional			

	Pass	Fail	N/A
Gauges and Indicators			
Load moment indicator operational			
Drum rotation indicator functioning			
Boom length indicator functioning			
Boom angle indicator functioning			
Engine: hydraulic, air, electrical, oil pressure, temperature, and fuel			
Operational Inspection			
Correct counterweight for the load			
Main hoist control functioning			
Auxiliary hoist control functioning			
Anti-two block in place and functioning			
Swing brake			
Lights and horns functional			

Notes:

Date_____

Operator		Signature	
Crane number		Model	

Visual Inspection	Pass	Fail	N/A		Pass	Fail	N/A
Engine fluid level correct (check dip stick or sight glass)				**Gauges and Indicators**			
Hydraulic fluid level correct (check dip stick or sight glass)				Load moment indicator operational			
Hydraulic system exhibits no apparent weeping or leaks				Drum rotation indicator functioning			
Air system exhibits no audible leaks				Boom length indicator functioning			
Tire pressure acceptable and tire not damaged				Boom angle indicator functioning			
Telescoping boom exhibits no damage to structure, wear pads, boom stops, or cylinder				Engine: hydraulic, air, electrical, oil pressure, temperature, and fuel			
Wire rope free of dirt, excess lube, kinks, and wires and spooled correctly				**Operational Inspection**			
Reeving correct				Correct counterweight for the load			
Wedge sockets and wire rope clips not distorted, cracked, or missing				Main hoist control functioning			
Block not damaged				Auxiliary hoist control functioning			
Ball and hook is free to swivel and rotate				Anti-two block in place and functioning			
Guards are in place				Swing brake			
Outrigger float(s) secured with pad pin				Lights and horns functional			
Cab							
Handrails in place and not damaged							
Operator's manual in vehicle							
Load chart legible and visible to operator							
Hand signal chart visible to workers							
Charged fire extinguisher in place							
Cab glass not cracked and wipers are functional							

Notes:

Date_____

Operator
Crane number

Signature	
Model	

Visual Inspection	Pass	Fail	N/A
Engine fluid level correct (check dip stick or sight glass)			
Hydraulic fluid level correct (check dip stick or sight glass)			
Hydraulic system exhibits no apparent weeping or leaks			
Air system exhibits no audible leaks			
Tire pressure acceptable and tire not damaged			
Telescoping boom exhibits no damage to structure, wear pads, boom stops, or cylinder			
Wire rope free of dirt, excess lube, kinks, and wires and spooled correctly			
Reeving correct			
Wedge sockets and wire rope clips not distorted, cracked, or missing			
Block not damaged			
Ball and hook is free to swivel and rotate			
Guards are in place			
Outrigger float(s) secured with pad pin			
Cab			
Handrails in place and not damaged			
Operator's manual in vehicle			
Load chart legible and visible to operator			
Hand signal chart visible to workers			
Charged fire extinguisher in place			
Cab glass not cracked and wipers are functional			

	Pass	Fail	N/A
Gauges and Indicators			
Load moment indicator operational			
Drum rotation indicator functioning			
Boom length indicator functioning			
Boom angle indicator functioning			
Engine: hydraulic, air, electrical, oil pressure, temperature, and fuel			
Operational Inspection			
Correct counterweight for the load			
Main hoist control functioning			
Auxiliary hoist control functioning			
Anti-two block in place and functioning			
Swing brake			
Lights and horns functional			

Notes:

Date_____

Operator		Signature	
Crane number		Model	

Visual Inspection	Pass	Fail	N/A		Pass	Fail	N/A
Engine fluid level correct (check dip stick or sight glass)				**Gauges and Indicators**			
Hydraulic fluid level correct (check dip stick or sight glass)				Load moment indicator operational			
Hydraulic system exhibits no apparent weeping or leaks				Drum rotation indicator functioning			
Air system exhibits no audible leaks				Boom length indicator functioning			
Tire pressure acceptable and tire not damaged				Boom angle indicator functioning			
Telescoping boom exhibits no damage to structure, wear pads, boom stops, or cylinder				Engine: hydraulic, air, electrical, oil pressure, temperature, and fuel			
Wire rope free of dirt, excess lube, kinks, and wires and spooled correctly				**Operational Inspection**			
Reeving correct				Correct counterweight for the load			
Wedge sockets and wire rope clips not distorted, cracked, or missing				Main hoist control functioning			
Block not damaged				Auxiliary hoist control functioning			
Ball and hook is free to swivel and rotate				Anti-two block in place and functioning			
Guards are in place				Swing brake			
Outrigger float(s) secured with pad pin				Lights and horns functional			
Cab							
Handrails in place and not damaged							
Operator's manual in vehicle							
Load chart legible and visible to operator							
Hand signal chart visible to workers							
Charged fire extinguisher in place							
Cab glass not cracked and wipers are functional							

Notes:

Date_____

Operator
Crane number

Signature	
Model	

Visual Inspection	Pass	Fail	N/A
Engine fluid level correct (check dip stick or sight glass)			
Hydraulic fluid level correct (check dip stick or sight glass)			
Hydraulic system exhibits no apparent weeping or leaks			
Air system exhibits no audible leaks			
Tire pressure acceptable and tire not damaged			
Telescoping boom exhibits no damage to structure, wear pads, boom stops, or cylinder			
Wire rope free of dirt, excess lube, kinks, and wires and spooled correctly			
Reeving correct			
Wedge sockets and wire rope clips not distorted, cracked, or missing			
Block not damaged			
Ball and hook is free to swivel and rotate			
Guards are in place			
Outrigger float(s) secured with pad pin			
Cab			
Handrails in place and not damaged			
Operator's manual in vehicle			
Load chart legible and visible to operator			
Hand signal chart visible to workers			
Charged fire extinguisher in place			
Cab glass not cracked and wipers are functional			

	Pass	Fail	N/A
Gauges and Indicators			
Load moment indicator operational			
Drum rotation indicator functioning			
Boom length indicator functioning			
Boom angle indicator functioning			
Engine: hydraulic, air, electrical, oil pressure, temperature, and fuel			
Operational Inspection			
Correct counterweight for the load			
Main hoist control functioning			
Auxiliary hoist control functioning			
Anti-two block in place and functioning			
Swing brake			
Lights and horns functional			

Notes:

Date_____

Operator		Signature	
Crane number		Model	

Visual Inspection	Pass	Fail	N/A		Pass	Fail	N/A
Engine fluid level correct (check dip stick or sight glass)				**Gauges and Indicators**			
Hydraulic fluid level correct (check dip stick or sight glass)				Load moment indicator operational			
Hydraulic system exhibits no apparent weeping or leaks				Drum rotation indicator functioning			
Air system exhibits no audible leaks				Boom length indicator functioning			
Tire pressure acceptable and tire not damaged				Boom angle indicator functioning			
Telescoping boom exhibits no damage to structure, wear pads, boom stops, or cylinder				Engine: hydraulic, air, electrical, oil pressure, temperature, and fuel			
Wire rope free of dirt, excess lube, kinks, and wires and spooled correctly				**Operational Inspection**			
Reeving correct				Correct counterweight for the load			
Wedge sockets and wire rope clips not distorted, cracked, or missing				Main hoist control functioning			
Block not damaged				Auxiliary hoist control functioning			
Ball and hook is free to swivel and rotate				Anti-two block in place and functioning			
Guards are in place				Swing brake			
Outrigger float(s) secured with pad pin				Lights and horns functional			
Cab							
Handrails in place and not damaged							
Operator's manual in vehicle							
Load chart legible and visible to operator							
Hand signal chart visible to workers							
Charged fire extinguisher in place							
Cab glass not cracked and wipers are functional							

Notes:

Date_____

Operator
Crane number

Signature
Model

Visual Inspection	Pass	Fail	N/A
Engine fluid level correct (check dip stick or sight glass)			
Hydraulic fluid level correct (check dip stick or sight glass)			
Hydraulic system exhibits no apparent weeping or leaks			
Air system exhibits no audible leaks			
Tire pressure acceptable and tire not damaged			
Telescoping boom exhibits no damage to structure, wear pads, boom stops, or cylinder			
Wire rope free of dirt, excess lube, kinks, and wires and spooled correctly			
Reeving correct			
Wedge sockets and wire rope clips not distorted, cracked, or missing			
Block not damaged			
Ball and hook is free to swivel and rotate			
Guards are in place			
Outrigger float(s) secured with pad pin			
Cab			
Handrails in place and not damaged			
Operator's manual in vehicle			
Load chart legible and visible to operator			
Hand signal chart visible to workers			
Charged fire extinguisher in place			
Cab glass not cracked and wipers are functional			

	Pass	Fail	N/A
Gauges and Indicators			
Load moment indicator operational			
Drum rotation indicator functioning			
Boom length indicator functioning			
Boom angle indicator functioning			
Engine: hydraulic, air, electrical, oil pressure, temperature, and fuel			
Operational Inspection			
Correct counterweight for the load			
Main hoist control functioning			
Auxiliary hoist control functioning			
Anti-two block in place and functioning			
Swing brake			
Lights and horns functional			

Notes:

Date_____

Operator		Signature		
Crane number		Model		

Visual Inspection	Pass	Fail	N/A		Pass	Fail	N/A
Engine fluid level correct (check dip stick or sight glass)				**Gauges and Indicators**			
Hydraulic fluid level correct (check dip stick or sight glass)				Load moment indicator operational			
Hydraulic system exhibits no apparent weeping or leaks				Drum rotation indicator functioning			
Air system exhibits no audible leaks				Boom length indicator functioning			
Tire pressure acceptable and tire not damaged				Boom angle indicator functioning			
Telescoping boom exhibits no damage to structure, wear pads, boom stops, or cylinder				Engine: hydraulic, air, electrical, oil pressure, temperature, and fuel			
Wire rope free of dirt, excess lube, kinks, and wires and spooled correctly				**Operational Inspection**			
Reeving correct				Correct counterweight for the load			
Wedge sockets and wire rope clips not distorted, cracked, or missing				Main hoist control functioning			
Block not damaged				Auxiliary hoist control functioning			
Ball and hook is free to swivel and rotate				Anti-two block in place and functioning			
Guards are in place				Swing brake			
Outrigger float(s) secured with pad pin				Lights and horns functional			
Cab							
Handrails in place and not damaged							
Operator's manual in vehicle							
Load chart legible and visible to operator							
Hand signal chart visible to workers							
Charged fire extinguisher in place							
Cab glass not cracked and wipers are functional							

Notes:

Date_____

Operator
Crane number

Signature	
Model	

Visual Inspection	Pass	Fail	N/A
Engine fluid level correct (check dip stick or sight glass)			
Hydraulic fluid level correct (check dip stick or sight glass)			
Hydraulic system exhibits no apparent weeping or leaks			
Air system exhibits no audible leaks			
Tire pressure acceptable and tire not damaged			
Telescoping boom exhibits no damage to structure, wear pads, boom stops, or cylinder			
Wire rope free of dirt, excess lube, kinks, and wires and spooled correctly			
Reeving correct			
Wedge sockets and wire rope clips not distorted, cracked, or missing			
Block not damaged			
Ball and hook is free to swivel and rotate			
Guards are in place			
Outrigger float(s) secured with pad pin			
Cab			
Handrails in place and not damaged			
Operator's manual in vehicle			
Load chart legible and visible to operator			
Hand signal chart visible to workers			
Charged fire extinguisher in place			
Cab glass not cracked and wipers are functional			

	Pass	Fail	N/A
Gauges and Indicators			
Load moment indicator operational			
Drum rotation indicator functioning			
Boom length indicator functioning			
Boom angle indicator functioning			
Engine: hydraulic, air, electrical, oil pressure, temperature, and fuel			
Operational Inspection			
Correct counterweight for the load			
Main hoist control functioning			
Auxiliary hoist control functioning			
Anti-two block in place and functioning			
Swing brake			
Lights and horns functional			

Notes:

Date_____

Operator		Signature	
Crane number		Model	

Visual Inspection	Pass	Fail	N/A		Pass	Fail	N/A
Engine fluid level correct (check dip stick or sight glass)				**Gauges and Indicators**			
Hydraulic fluid level correct (check dip stick or sight glass)				Load moment indicator operational			
Hydraulic system exhibits no apparent weeping or leaks				Drum rotation indicator functioning			
Air system exhibits no audible leaks				Boom length indicator functioning			
Tire pressure acceptable and tire not damaged				Boom angle indicator functioning			
Telescoping boom exhibits no damage to structure, wear pads, boom stops, or cylinder				Engine: hydraulic, air, electrical, oil pressure, temperature, and fuel			
Wire rope free of dirt, excess lube, kinks, and wires and spooled correctly				**Operational Inspection**			
Reeving correct				Correct counterweight for the load			
Wedge sockets and wire rope clips not distorted, cracked, or missing				Main hoist control functioning			
Block not damaged				Auxiliary hoist control functioning			
Ball and hook is free to swivel and rotate				Anti-two block in place and functioning			
Guards are in place				Swing brake			
Outrigger float(s) secured with pad pin				Lights and horns functional			
Cab							
Handrails in place and not damaged							
Operator's manual in vehicle							
Load chart legible and visible to operator							
Hand signal chart visible to workers							
Charged fire extinguisher in place							
Cab glass not cracked and wipers are functional							

Notes:

Date_____

Operator
Crane number

Signature	
Model	

Visual Inspection	Pass	Fail	N/A
Engine fluid level correct (check dip stick or sight glass)			
Hydraulic fluid level correct (check dip stick or sight glass)			
Hydraulic system exhibits no apparent weeping or leaks			
Air system exhibits no audible leaks			
Tire pressure acceptable and tire not damaged			
Telescoping boom exhibits no damage to structure, wear pads, boom stops, or cylinder			
Wire rope free of dirt, excess lube, kinks, and wires and spooled correctly			
Reeving correct			
Wedge sockets and wire rope clips not distorted, cracked, or missing			
Block not damaged			
Ball and hook is free to swivel and rotate			
Guards are in place			
Outrigger float(s) secured with pad pin			
Cab			
Handrails in place and not damaged			
Operator's manual in vehicle			
Load chart legible and visible to operator			
Hand signal chart visible to workers			
Charged fire extinguisher in place			
Cab glass not cracked and wipers are functional			

	Pass	Fail	N/A
Gauges and Indicators			
Load moment indicator operational			
Drum rotation indicator functioning			
Boom length indicator functioning			
Boom angle indicator functioning			
Engine: hydraulic, air, electrical, oil pressure, temperature, and fuel			
Operational Inspection			
Correct counterweight for the load			
Main hoist control functioning			
Auxiliary hoist control functioning			
Anti-two block in place and functioning			
Swing brake			
Lights and horns functional			

Notes:

Date _____

Operator
Crane number

Signature	
Model	

Visual Inspection	Pass	Fail	N/A
Engine fluid level correct (check dip stick or sight glass)			
Hydraulic fluid level correct (check dip stick or sight glass)			
Hydraulic system exhibits no apparent weeping or leaks			
Air system exhibits no audible leaks			
Tire pressure acceptable and tire not damaged			
Telescoping boom exhibits no damage to structure, wear pads, boom stops, or cylinder			
Wire rope free of dirt, excess lube, kinks, and wires and spooled correctly			
Reeving correct			
Wedge sockets and wire rope clips not distorted, cracked, or missing			
Block not damaged			
Ball and hook is free to swivel and rotate			
Guards are in place			
Outrigger float(s) secured with pad pin			
Cab			
Handrails in place and not damaged			
Operator's manual in vehicle			
Load chart legible and visible to operator			
Hand signal chart visible to workers			
Charged fire extinguisher in place			
Cab glass not cracked and wipers are functional			

	Pass	Fail	N/A
Gauges and Indicators			
Load moment indicator operational			
Drum rotation indicator functioning			
Boom length indicator functioning			
Boom angle indicator functioning			
Engine: hydraulic, air, electrical, oil pressure, temperature, and fuel			
Operational Inspection			
Correct counterweight for the load			
Main hoist control functioning			
Auxiliary hoist control functioning			
Anti-two block in place and functioning			
Swing brake			
Lights and horns functional			

Notes:

Date_____

Operator
Crane number

Signature	
Model	

Visual Inspection	Pass	Fail	N/A
Engine fluid level correct (check dip stick or sight glass)			
Hydraulic fluid level correct (check dip stick or sight glass)			
Hydraulic system exhibits no apparent weeping or leaks			
Air system exhibits no audible leaks			
Tire pressure acceptable and tire not damaged			
Telescoping boom exhibits no damage to structure, wear pads, boom stops, or cylinder			
Wire rope free of dirt, excess lube, kinks, and wires and spooled correctly			
Reeving correct			
Wedge sockets and wire rope clips not distorted, cracked, or missing			
Block not damaged			
Ball and hook is free to swivel and rotate			
Guards are in place			
Outrigger float(s) secured with pad pin			
Cab			
Handrails in place and not damaged			
Operator's manual in vehicle			
Load chart legible and visible to operator			
Hand signal chart visible to workers			
Charged fire extinguisher in place			
Cab glass not cracked and wipers are functional			

	Pass	Fail	N/A
Gauges and Indicators			
Load moment indicator operational			
Drum rotation indicator functioning			
Boom length indicator functioning			
Boom angle indicator functioning			
Engine: hydraulic, air, electrical, oil pressure, temperature, and fuel			
Operational Inspection			
Correct counterweight for the load			
Main hoist control functioning			
Auxiliary hoist control functioning			
Anti-two block in place and functioning			
Swing brake			
Lights and horns functional			

Notes:

Date_____

Operator		Signature	
Crane number		Model	

Visual Inspection	Pass	Fail	N/A
Engine fluid level correct (check dip stick or sight glass)			
Hydraulic fluid level correct (check dip stick or sight glass)			
Hydraulic system exhibits no apparent weeping or leaks			
Air system exhibits no audible leaks			
Tire pressure acceptable and tire not damaged			
Telescoping boom exhibits no damage to structure, wear pads, boom stops, or cylinder			
Wire rope free of dirt, excess lube, kinks, and wires and spooled correctly			
Reeving correct			
Wedge sockets and wire rope clips not distorted, cracked, or missing			
Block not damaged			
Ball and hook is free to swivel and rotate			
Guards are in place			
Outrigger float(s) secured with pad pin			
Cab			
Handrails in place and not damaged			
Operator's manual in vehicle			
Load chart legible and visible to operator			
Hand signal chart visible to workers			
Charged fire extinguisher in place			
Cab glass not cracked and wipers are functional			

	Pass	Fail	N/A
Gauges and Indicators			
Load moment indicator operational			
Drum rotation indicator functioning			
Boom length indicator functioning			
Boom angle indicator functioning			
Engine: hydraulic, air, electrical, oil pressure, temperature, and fuel			
Operational Inspection			
Correct counterweight for the load			
Main hoist control functioning			
Auxiliary hoist control functioning			
Anti-two block in place and functioning			
Swing brake			
Lights and horns functional			

Notes:

Date_____

Operator
Crane number

Signature	
Model	

Visual Inspection	Pass	Fail	N/A
Engine fluid level correct (check dip stick or sight glass)			
Hydraulic fluid level correct (check dip stick or sight glass)			
Hydraulic system exhibits no apparent weeping or leaks			
Air system exhibits no audible leaks			
Tire pressure acceptable and tire not damaged			
Telescoping boom exhibits no damage to structure, wear pads, boom stops, or cylinder			
Wire rope free of dirt, excess lube, kinks, and wires and spooled correctly			
Reeving correct			
Wedge sockets and wire rope clips not distorted, cracked, or missing			
Block not damaged			
Ball and hook is free to swivel and rotate			
Guards are in place			
Outrigger float(s) secured with pad pin			
Cab			
Handrails in place and not damaged			
Operator's manual in vehicle			
Load chart legible and visible to operator			
Hand signal chart visible to workers			
Charged fire extinguisher in place			
Cab glass not cracked and wipers are functional			

	Pass	Fail	N/A
Gauges and Indicators			
Load moment indicator operational			
Drum rotation indicator functioning			
Boom length indicator functioning			
Boom angle indicator functioning			
Engine: hydraulic, air, electrical, oil pressure, temperature, and fuel			
Operational Inspection			
Correct counterweight for the load			
Main hoist control functioning			
Auxiliary hoist control functioning			
Anti-two block in place and functioning			
Swing brake			
Lights and horns functional			

Notes:

Date_____

Operator		Signature	
Crane number		Model	

Visual Inspection	Pass	Fail	N/A		Pass	Fail	N/A
Engine fluid level correct (check dip stick or sight glass)				**Gauges and Indicators**			
Hydraulic fluid level correct (check dip stick or sight glass)				Load moment indicator operational			
Hydraulic system exhibits no apparent weeping or leaks				Drum rotation indicator functioning			
Air system exhibits no audible leaks				Boom length indicator functioning			
Tire pressure acceptable and tire not damaged				Boom angle indicator functioning			
Telescoping boom exhibits no damage to structure, wear pads, boom stops, or cylinder				Engine: hydraulic, air, electrical, oil pressure, temperature, and fuel			
Wire rope free of dirt, excess lube, kinks, and wires and spooled correctly				**Operational Inspection**			
Reeving correct				Correct counterweight for the load			
Wedge sockets and wire rope clips not distorted, cracked, or missing				Main hoist control functioning			
Block not damaged				Auxiliary hoist control functioning			
Ball and hook is free to swivel and rotate				Anti-two block in place and functioning			
Guards are in place				Swing brake			
Outrigger float(s) secured with pad pin				Lights and horns functional			
Cab							
Handrails in place and not damaged							
Operator's manual in vehicle							
Load chart legible and visible to operator							
Hand signal chart visible to workers							
Charged fire extinguisher in place							
Cab glass not cracked and wipers are functional							

Notes:

Date_____

Operator
Crane number

Signature	
Model	

Visual Inspection	Pass	Fail	N/A
Engine fluid level correct (check dip stick or sight glass)			
Hydraulic fluid level correct (check dip stick or sight glass)			
Hydraulic system exhibits no apparent weeping or leaks			
Air system exhibits no audible leaks			
Tire pressure acceptable and tire not damaged			
Telescoping boom exhibits no damage to structure, wear pads, boom stops, or cylinder			
Wire rope free of dirt, excess lube, kinks, and wires and spooled correctly			
Reeving correct			
Wedge sockets and wire rope clips not distorted, cracked, or missing			
Block not damaged			
Ball and hook is free to swivel and rotate			
Guards are in place			
Outrigger float(s) secured with pad pin			
Cab			
Handrails in place and not damaged			
Operator's manual in vehicle			
Load chart legible and visible to operator			
Hand signal chart visible to workers			
Charged fire extinguisher in place			
Cab glass not cracked and wipers are functional			

	Pass	Fail	N/A
Gauges and Indicators			
Load moment indicator operational			
Drum rotation indicator functioning			
Boom length indicator functioning			
Boom angle indicator functioning			
Engine: hydraulic, air, electrical, oil pressure, temperature, and fuel			
Operational Inspection			
Correct counterweight for the load			
Main hoist control functioning			
Auxiliary hoist control functioning			
Anti-two block in place and functioning			
Swing brake			
Lights and horns functional			

Notes:

Date_____

Operator		Signature	
Crane number		Model	

Visual Inspection	Pass	Fail	N/A		Pass	Fail	N/A
Engine fluid level correct (check dip stick or sight glass)				**Gauges and Indicators**			
Hydraulic fluid level correct (check dip stick or sight glass)				Load moment indicator operational			
Hydraulic system exhibits no apparent weeping or leaks				Drum rotation indicator functioning			
Air system exhibits no audible leaks				Boom length indicator functioning			
Tire pressure acceptable and tire not damaged				Boom angle indicator functioning			
Telescoping boom exhibits no damage to structure, wear pads, boom stops, or cylinder				Engine: hydraulic, air, electrical, oil pressure, temperature, and fuel			
Wire rope free of dirt, excess lube, kinks, and wires and spooled correctly				**Operational Inspection**			
Reeving correct				Correct counterweight for the load			
Wedge sockets and wire rope clips not distorted, cracked, or missing				Main hoist control functioning			
Block not damaged				Auxiliary hoist control functioning			
Ball and hook is free to swivel and rotate				Anti-two block in place and functioning			
Guards are in place				Swing brake			
Outrigger float(s) secured with pad pin				Lights and horns functional			
Cab							
Handrails in place and not damaged							
Operator's manual in vehicle							
Load chart legible and visible to operator							
Hand signal chart visible to workers							
Charged fire extinguisher in place							
Cab glass not cracked and wipers are functional							

Notes:

Date_____

Operator		Signature	
Crane number		Model	

Visual Inspection	Pass	Fail	N/A
Engine fluid level correct (check dip stick or sight glass)			
Hydraulic fluid level correct (check dip stick or sight glass)			
Hydraulic system exhibits no apparent weeping or leaks			
Air system exhibits no audible leaks			
Tire pressure acceptable and tire not damaged			
Telescoping boom exhibits no damage to structure, wear pads, boom stops, or cylinder			
Wire rope free of dirt, excess lube, kinks, and wires and spooled correctly			
Reeving correct			
Wedge sockets and wire rope clips not distorted, cracked, or missing			
Block not damaged			
Ball and hook is free to swivel and rotate			
Guards are in place			
Outrigger float(s) secured with pad pin			
Cab			
Handrails in place and not damaged			
Operator's manual in vehicle			
Load chart legible and visible to operator			
Hand signal chart visible to workers			
Charged fire extinguisher in place			
Cab glass not cracked and wipers are functional			

	Pass	Fail	N/A
Gauges and Indicators			
Load moment indicator operational			
Drum rotation indicator functioning			
Boom length indicator functioning			
Boom angle indicator functioning			
Engine: hydraulic, air, electrical, oil pressure, temperature, and fuel			
Operational Inspection			
Correct counterweight for the load			
Main hoist control functioning			
Auxiliary hoist control functioning			
Anti-two block in place and functioning			
Swing brake			
Lights and horns functional			

Notes:

Date_____

Operator		Signature	
Crane number		Model	

Visual Inspection	Pass	Fail	N/A
Engine fluid level correct (check dip stick or sight glass)			
Hydraulic fluid level correct (check dip stick or sight glass)			
Hydraulic system exhibits no apparent weeping or leaks			
Air system exhibits no audible leaks			
Tire pressure acceptable and tire not damaged			
Telescoping boom exhibits no damage to structure, wear pads, boom stops, or cylinder			
Wire rope free of dirt, excess lube, kinks, and wires and spooled correctly			
Reeving correct			
Wedge sockets and wire rope clips not distorted, cracked, or missing			
Block not damaged			
Ball and hook is free to swivel and rotate			
Guards are in place			
Outrigger float(s) secured with pad pin			
Cab			
Handrails in place and not damaged			
Operator's manual in vehicle			
Load chart legible and visible to operator			
Hand signal chart visible to workers			
Charged fire extinguisher in place			
Cab glass not cracked and wipers are functional			

	Pass	Fail	N/A
Gauges and Indicators			
Load moment indicator operational			
Drum rotation indicator functioning			
Boom length indicator functioning			
Boom angle indicator functioning			
Engine: hydraulic, air, electrical, oil pressure, temperature, and fuel			
Operational Inspection			
Correct counterweight for the load			
Main hoist control functioning			
Auxiliary hoist control functioning			
Anti-two block in place and functioning			
Swing brake			
Lights and horns functional			

Notes:

Date_____

Operator		Signature	
Crane number		Model	

Visual Inspection	Pass	Fail	N/A		Pass	Fail	N/A
Engine fluid level correct (check dip stick or sight glass)				**Gauges and Indicators**			
Hydraulic fluid level correct (check dip stick or sight glass)				Load moment indicator operational			
Hydraulic system exhibits no apparent weeping or leaks				Drum rotation indicator functioning			
Air system exhibits no audible leaks				Boom length indicator functioning			
Tire pressure acceptable and tire not damaged				Boom angle indicator functioning			
Telescoping boom exhibits no damage to structure, wear pads, boom stops, or cylinder				Engine: hydraulic, air, electrical, oil pressure, temperature, and fuel			
Wire rope free of dirt, excess lube, kinks, and wires and spooled correctly				**Operational Inspection**			
Reeving correct				Correct counterweight for the load			
Wedge sockets and wire rope clips not distorted, cracked, or missing				Main hoist control functioning			
Block not damaged				Auxiliary hoist control functioning			
Ball and hook is free to swivel and rotate				Anti-two block in place and functioning			
Guards are in place				Swing brake			
Outrigger float(s) secured with pad pin				Lights and horns functional			
Cab							
Handrails in place and not damaged							
Operator's manual in vehicle							
Load chart legible and visible to operator							
Hand signal chart visible to workers							
Charged fire extinguisher in place							
Cab glass not cracked and wipers are functional							

Notes:

ate_____

Operator		Signature	
Crane number		Model	

Visual Inspection	Pass	Fail	N/A		Pass	Fail	N/A
Engine fluid level correct (check dip stick or sight glass)				**Gauges and Indicators**			
Hydraulic fluid level correct (check dip stick or sight glass)				Load moment indicator operational			
Hydraulic system exhibits no apparent weeping or leaks				Drum rotation indicator functioning			
Air system exhibits no audible leaks				Boom length indicator functioning			
Tire pressure acceptable and tire not damaged				Boom angle indicator functioning			
Telescoping boom exhibits no damage to structure, wear pads, boom stops, or cylinder				Engine: hydraulic, air, electrical, oil pressure, temperature, and fuel			
Wire rope free of dirt, excess lube, kinks, and wires and spooled correctly				**Operational Inspection**			
Reeving correct				Correct counterweight for the load			
Wedge sockets and wire rope clips not distorted, cracked, or missing				Main hoist control functioning			
Block not damaged				Auxiliary hoist control functioning			
Ball and hook is free to swivel and rotate				Anti-two block in place and functioning			
Guards are in place				Swing brake			
Outrigger float(s) secured with pad pin				Lights and horns functional			
Cab							
Handrails in place and not damaged							
Operator's manual in vehicle							
Load chart legible and visible to operator							
Hand signal chart visible to workers							
Charged fire extinguisher in place							
Cab glass not cracked and wipers are functional							

otes:

Date_____

Operator
Crane number

Signature	
Model	

Visual Inspection	Pass	Fail	N/A
Engine fluid level correct (check dip stick or sight glass)			
Hydraulic fluid level correct (check dip stick or sight glass)			
Hydraulic system exhibits no apparent weeping or leaks			
Air system exhibits no audible leaks			
Tire pressure acceptable and tire not damaged			
Telescoping boom exhibits no damage to structure, wear pads, boom stops, or cylinder			
Wire rope free of dirt, excess lube, kinks, and wires and spooled correctly			
Reeving correct			
Wedge sockets and wire rope clips not distorted, cracked, or missing			
Block not damaged			
Ball and hook is free to swivel and rotate			
Guards are in place			
Outrigger float(s) secured with pad pin			
Cab			
Handrails in place and not damaged			
Operator's manual in vehicle			
Load chart legible and visible to operator			
Hand signal chart visible to workers			
Charged fire extinguisher in place			
Cab glass not cracked and wipers are functional			

	Pass	Fail	N/A
Gauges and Indicators			
Load moment indicator operational			
Drum rotation indicator functioning			
Boom length indicator functioning			
Boom angle indicator functioning			
Engine: hydraulic, air, electrical, oil pressure, temperature, and fuel			
Operational Inspection			
Correct counterweight for the load			
Main hoist control functioning			
Auxiliary hoist control functioning			
Anti-two block in place and functioning			
Swing brake			
Lights and horns functional			

Notes:

Date _____

Operator		Signature	
Crane number		Model	

Visual Inspection	Pass	Fail	N/A		Pass	Fail	N/A
Engine fluid level correct (check dip stick or sight glass)				Gauges and Indicators			
Hydraulic fluid level correct (check dip stick or sight glass)				Load moment indicator operational			
Hydraulic system exhibits no apparent weeping or leaks				Drum rotation indicator functioning			
Air system exhibits no audible leaks				Boom length indicator functioning			
Tire pressure acceptable and tire not damaged				Boom angle indicator functioning			
Telescoping boom exhibits no damage to structure, wear pads, boom stops, or cylinder				Engine: hydraulic, air, electrical, oil pressure, temperature, and fuel			
Wire rope free of dirt, excess lube, kinks, and wires and spooled correctly				Operational Inspection			
Reeving correct				Correct counterweight for the load			
Wedge sockets and wire rope clips not distorted, cracked, or missing				Main hoist control functioning			
Block not damaged				Auxiliary hoist control functioning			
Ball and hook is free to swivel and rotate				Anti-two block in place and functioning			
Guards are in place				Swing brake			
Outrigger float(s) secured with pad pin				Lights and horns functional			
Cab							
Handrails in place and not damaged							
Operator's manual in vehicle							
Load chart legible and visible to operator							
Hand signal chart visible to workers							
Charged fire extinguisher in place							
Cab glass not cracked and wipers are functional							

Notes:

Date_____

Operator
Crane number

Signature	
Model	

Visual Inspection	Pass	Fail	N/A
Engine fluid level correct (check dip stick or sight glass)			
Hydraulic fluid level correct (check dip stick or sight glass)			
Hydraulic system exhibits no apparent weeping or leaks			
Air system exhibits no audible leaks			
Tire pressure acceptable and tire not damaged			
Telescoping boom exhibits no damage to structure, wear pads, boom stops, or cylinder			
Wire rope free of dirt, excess lube, kinks, and wires and spooled correctly			
Reeving correct			
Wedge sockets and wire rope clips not distorted, cracked, or missing			
Block not damaged			
Ball and hook is free to swivel and rotate			
Guards are in place			
Outrigger float(s) secured with pad pin			
Cab			
Handrails in place and not damaged			
Operator's manual in vehicle			
Load chart legible and visible to operator			
Hand signal chart visible to workers			
Charged fire extinguisher in place			
Cab glass not cracked and wipers are functional			

	Pass	Fail	N/A
Gauges and Indicators			
Load moment indicator operational			
Drum rotation indicator functioning			
Boom length indicator functioning			
Boom angle indicator functioning			
Engine: hydraulic, air, electrical, oil pressure, temperature, and fuel			
Operational Inspection			
Correct counterweight for the load			
Main hoist control functioning			
Auxiliary hoist control functioning			
Anti-two block in place and functioning			
Swing brake			
Lights and horns functional			

Notes:

Date_____

Operator		Signature	
Crane number		Model	

Visual Inspection	Pass	Fail	N/A		Pass	Fail	N/A
Engine fluid level correct (check dip stick or sight glass)				**Gauges and Indicators**			
Hydraulic fluid level correct (check dip stick or sight glass)				Load moment indicator operational			
Hydraulic system exhibits no apparent weeping or leaks				Drum rotation indicator functioning			
Air system exhibits no audible leaks				Boom length indicator functioning			
Tire pressure acceptable and tire not damaged				Boom angle indicator functioning			
Telescoping boom exhibits no damage to structure, wear pads, boom stops, or cylinder				Engine: hydraulic, air, electrical, oil pressure, temperature, and fuel			
Wire rope free of dirt, excess lube, kinks, and wires and spooled correctly				**Operational Inspection**			
Reeving correct				Correct counterweight for the load			
Wedge sockets and wire rope clips not distorted, cracked, or missing				Main hoist control functioning			
Block not damaged				Auxiliary hoist control functioning			
Ball and hook is free to swivel and rotate				Anti-two block in place and functioning			
Guards are in place				Swing brake			
Outrigger float(s) secured with pad pin				Lights and horns functional			
Cab							
Handrails in place and not damaged							
Operator's manual in vehicle							
Load chart legible and visible to operator							
Hand signal chart visible to workers							
Charged fire extinguisher in place							
Cab glass not cracked and wipers are functional							

Notes:

Date_____

Operator
Crane number

Signature	
Model	

Visual Inspection	Pass	Fail	N/A
Engine fluid level correct (check dip stick or sight glass)			
Hydraulic fluid level correct (check dip stick or sight glass)			
Hydraulic system exhibits no apparent weeping or leaks			
Air system exhibits no audible leaks			
Tire pressure acceptable and tire not damaged			
Telescoping boom exhibits no damage to structure, wear pads, boom stops, or cylinder			
Wire rope free of dirt, excess lube, kinks, and wires and spooled correctly			
Reeving correct			
Wedge sockets and wire rope clips not distorted, cracked, or missing			
Block not damaged			
Ball and hook is free to swivel and rotate			
Guards are in place			
Outrigger float(s) secured with pad pin			
Cab			
Handrails in place and not damaged			
Operator's manual in vehicle			
Load chart legible and visible to operator			
Hand signal chart visible to workers			
Charged fire extinguisher in place			
Cab glass not cracked and wipers are functional			

	Pass	Fail	N/A
Gauges and Indicators			
Load moment indicator operational			
Drum rotation indicator functioning			
Boom length indicator functioning			
Boom angle indicator functioning			
Engine: hydraulic, air, electrical, oil pressure, temperature, and fuel			
Operational Inspection			
Correct counterweight for the load			
Main hoist control functioning			
Auxiliary hoist control functioning			
Anti-two block in place and functioning			
Swing brake			
Lights and horns functional			

Notes:

Date_____

Operator		Signature	
Crane number		Model	

Visual Inspection	Pass	Fail	N/A		Pass	Fail	N/A
Engine fluid level correct (check dip stick or sight glass)				**Gauges and Indicators**			
Hydraulic fluid level correct (check dip stick or sight glass)				Load moment indicator operational			
Hydraulic system exhibits no apparent weeping or leaks				Drum rotation indicator functioning			
Air system exhibits no audible leaks				Boom length indicator functioning			
Tire pressure acceptable and tire not damaged				Boom angle indicator functioning			
Telescoping boom exhibits no damage to structure, wear pads, boom stops, or cylinder				Engine: hydraulic, air, electrical, oil pressure, temperature, and fuel			
Wire rope free of dirt, excess lube, kinks, and wires and spooled correctly				**Operational Inspection**			
Reeving correct				Correct counterweight for the load			
Wedge sockets and wire rope clips not distorted, cracked, or missing				Main hoist control functioning			
Block not damaged				Auxiliary hoist control functioning			
Ball and hook is free to swivel and rotate				Anti-two block in place and functioning			
Guards are in place				Swing brake			
Outrigger float(s) secured with pad pin				Lights and horns functional			
Cab							
Handrails in place and not damaged							
Operator's manual in vehicle							
Load chart legible and visible to operator							
Hand signal chart visible to workers							
Charged fire extinguisher in place							
Cab glass not cracked and wipers are functional							

Notes:

Date_____

Operator
Crane number

Signature
Model

Visual Inspection	Pass	Fail	N/A
Engine fluid level correct (check dip stick or sight glass)			
Hydraulic fluid level correct (check dip stick or sight glass)			
Hydraulic system exhibits no apparent weeping or leaks			
Air system exhibits no audible leaks			
Tire pressure acceptable and tire not damaged			
Telescoping boom exhibits no damage to structure, wear pads, boom stops, or cylinder			
Wire rope free of dirt, excess lube, kinks, and wires and spooled correctly			
Reeving correct			
Wedge sockets and wire rope clips not distorted, cracked, or missing			
Block not damaged			
Ball and hook is free to swivel and rotate			
Guards are in place			
Outrigger float(s) secured with pad pin			
Cab			
Handrails in place and not damaged			
Operator's manual in vehicle			
Load chart legible and visible to operator			
Hand signal chart visible to workers			
Charged fire extinguisher in place			
Cab glass not cracked and wipers are functional			

	Pass	Fail	N/A
Gauges and Indicators			
Load moment indicator operational			
Drum rotation indicator functioning			
Boom length indicator functioning			
Boom angle indicator functioning			
Engine: hydraulic, air, electrical, oil pressure, temperature, and fuel			
Operational Inspection			
Correct counterweight for the load			
Main hoist control functioning			
Auxiliary hoist control functioning			
Anti-two block in place and functioning			
Swing brake			
Lights and horns functional			

Notes:

Date_____

Operator		Signature	
Crane number		Model	

Visual Inspection	Pass	Fail	N/A		Pass	Fail	N/A
Engine fluid level correct (check dip stick or sight glass)				**Gauges and Indicators**			
Hydraulic fluid level correct (check dip stick or sight glass)				Load moment indicator operational			
Hydraulic system exhibits no apparent weeping or leaks				Drum rotation indicator functioning			
Air system exhibits no audible leaks				Boom length indicator functioning			
Tire pressure acceptable and tire not damaged				Boom angle indicator functioning			
Telescoping boom exhibits no damage to structure, wear pads, boom stops, or cylinder				Engine: hydraulic, air, electrical, oil pressure, temperature, and fuel			
Wire rope free of dirt, excess lube, kinks, and wires and spooled correctly				**Operational Inspection**			
Reeving correct				Correct counterweight for the load			
Wedge sockets and wire rope clips not distorted, cracked, or missing				Main hoist control functioning			
Block not damaged				Auxiliary hoist control functioning			
Ball and hook is free to swivel and rotate				Anti-two block in place and functioning			
Guards are in place				Swing brake			
Outrigger float(s) secured with pad pin				Lights and horns functional			
Cab							
Handrails in place and not damaged							
Operator's manual in vehicle							
Load chart legible and visible to operator							
Hand signal chart visible to workers							
Charged fire extinguisher in place							
Cab glass not cracked and wipers are functional							

Notes:

Date_____

Operator		Signature	
Crane number		Model	

Visual Inspection	Pass	Fail	N/A	Gauges and Indicators	Pass	Fail	N/A
Engine fluid level correct (check dip stick or sight glass)				Load moment indicator operational			
Hydraulic fluid level correct (check dip stick or sight glass)				Drum rotation indicator functioning			
Hydraulic system exhibits no apparent weeping or leaks				Boom length indicator functioning			
Air system exhibits no audible leaks				Boom angle indicator functioning			
Tire pressure acceptable and tire not damaged				Engine: hydraulic, air, electrical, oil pressure, temperature, and fuel			
Telescoping boom exhibits no damage to structure, wear pads, boom stops, or cylinder				**Operational Inspection**			
Wire rope free of dirt, excess lube, kinks, and wires and spooled correctly				Correct counterweight for the load			
Reeving correct				Main hoist control functioning			
Wedge sockets and wire rope clips not distorted, cracked, or missing				Auxiliary hoist control functioning			
Block not damaged				Anti-two block in place and functioning			
Ball and hook is free to swivel and rotate				Swing brake			
Guards are in place				Lights and horns functional			
Outrigger float(s) secured with pad pin							
Cab							
Handrails in place and not damaged							
Operator's manual in vehicle							
Load chart legible and visible to operator							
Hand signal chart visible to workers							
Charged fire extinguisher in place							
Cab glass not cracked and wipers are functional							

Notes:

Date_____

Operator		Signature	
Crane number		Model	

Visual Inspection	Pass	Fail	N/A		Pass	Fail	N/A
Engine fluid level correct (check dip stick or sight glass)				**Gauges and Indicators**			
Hydraulic fluid level correct (check dip stick or sight glass)				Load moment indicator operational			
Hydraulic system exhibits no apparent weeping or leaks				Drum rotation indicator functioning			
Air system exhibits no audible leaks				Boom length indicator functioning			
Tire pressure acceptable and tire not damaged				Boom angle indicator functioning			
Telescoping boom exhibits no damage to structure, wear pads, boom stops, or cylinder				Engine: hydraulic, air, electrical, oil pressure, temperature, and fuel			
Wire rope free of dirt, excess lube, kinks, and wires and spooled correctly				**Operational Inspection**			
Reeving correct				Correct counterweight for the load			
Wedge sockets and wire rope clips not distorted, cracked, or missing				Main hoist control functioning			
Block not damaged				Auxiliary hoist control functioning			
Ball and hook is free to swivel and rotate				Anti-two block in place and functioning			
Guards are in place				Swing brake			
Outrigger float(s) secured with pad pin				Lights and horns functional			
Cab							
Handrails in place and not damaged							
Operator's manual in vehicle							
Load chart legible and visible to operator							
Hand signal chart visible to workers							
Charged fire extinguisher in place							
Cab glass not cracked and wipers are functional							

Notes:

Date_____

Operator
Crane number

Signature	
Model	

Visual Inspection	Pass	Fail	N/A
Engine fluid level correct (check dip stick or sight glass)			
Hydraulic fluid level correct (check dip stick or sight glass)			
Hydraulic system exhibits no apparent weeping or leaks			
Air system exhibits no audible leaks			
Tire pressure acceptable and tire not damaged			
Telescoping boom exhibits no damage to structure, wear pads, boom stops, or cylinder			
Wire rope free of dirt, excess lube, kinks, and wires and spooled correctly			
Reeving correct			
Wedge sockets and wire rope clips not distorted, cracked, or missing			
Block not damaged			
Ball and hook is free to swivel and rotate			
Guards are in place			
Outrigger float(s) secured with pad pin			
Cab			
Handrails in place and not damaged			
Operator's manual in vehicle			
Load chart legible and visible to operator			
Hand signal chart visible to workers			
Charged fire extinguisher in place			
Cab glass not cracked and wipers are functional			

	Pass	Fail	N/A
Gauges and Indicators			
Load moment indicator operational			
Drum rotation indicator functioning			
Boom length indicator functioning			
Boom angle indicator functioning			
Engine: hydraulic, air, electrical, oil pressure, temperature, and fuel			
Operational Inspection			
Correct counterweight for the load			
Main hoist control functioning			
Auxiliary hoist control functioning			
Anti-two block in place and functioning			
Swing brake			
Lights and horns functional			

Notes:

Date_____

Operator		Signature	
Crane number		Model	

Visual Inspection	Pass	Fail	N/A
Engine fluid level correct (check dip stick or sight glass)			
Hydraulic fluid level correct (check dip stick or sight glass)			
Hydraulic system exhibits no apparent weeping or leaks			
Air system exhibits no audible leaks			
Tire pressure acceptable and tire not damaged			
Telescoping boom exhibits no damage to structure, wear pads, boom stops, or cylinder			
Wire rope free of dirt, excess lube, kinks, and wires and spooled correctly			
Reeving correct			
Wedge sockets and wire rope clips not distorted, cracked, or missing			
Block not damaged			
Ball and hook is free to swivel and rotate			
Guards are in place			
Outrigger float(s) secured with pad pin			
Cab			
Handrails in place and not damaged			
Operator's manual in vehicle			
Load chart legible and visible to operator			
Hand signal chart visible to workers			
Charged fire extinguisher in place			
Cab glass not cracked and wipers are functional			

	Pass	Fail	N/A
Gauges and Indicators			
Load moment indicator operational			
Drum rotation indicator functioning			
Boom length indicator functioning			
Boom angle indicator functioning			
Engine: hydraulic, air, electrical, oil pressure, temperature, and fuel			
Operational Inspection			
Correct counterweight for the load			
Main hoist control functioning			
Auxiliary hoist control functioning			
Anti-two block in place and functioning			
Swing brake			
Lights and horns functional			

Notes:

Date_____

Operator
Crane number

Signature	
Model	

Visual Inspection	Pass	Fail	N/A
Engine fluid level correct (check dip stick or sight glass)			
Hydraulic fluid level correct (check dip stick or sight glass)			
Hydraulic system exhibits no apparent weeping or leaks			
Air system exhibits no audible leaks			
Tire pressure acceptable and tire not damaged			
Telescoping boom exhibits no damage to structure, wear pads, boom stops, or cylinder			
Wire rope free of dirt, excess lube, kinks, and wires and spooled correctly			
Reeving correct			
Wedge sockets and wire rope clips not distorted, cracked, or missing			
Block not damaged			
Ball and hook is free to swivel and rotate			
Guards are in place			
Outrigger float(s) secured with pad pin			
Cab			
Handrails in place and not damaged			
Operator's manual in vehicle			
Load chart legible and visible to operator			
Hand signal chart visible to workers			
Charged fire extinguisher in place			
Cab glass not cracked and wipers are functional			

	Pass	Fail	N/A
Gauges and Indicators			
Load moment indicator operational			
Drum rotation indicator functioning			
Boom length indicator functioning			
Boom angle indicator functioning			
Engine: hydraulic, air, electrical, oil pressure, temperature, and fuel			
Operational Inspection			
Correct counterweight for the load			
Main hoist control functioning			
Auxiliary hoist control functioning			
Anti-two block in place and functioning			
Swing brake			
Lights and horns functional			

Notes:

Date_____

Operator		Signature	
Crane number		Model	

Visual Inspection	Pass	Fail	N/A		Pass	Fail	N/A
Engine fluid level correct (check dip stick or sight glass)				Gauges and Indicators			
Hydraulic fluid level correct (check dip stick or sight glass)				Load moment indicator operational			
Hydraulic system exhibits no apparent weeping or leaks				Drum rotation indicator functioning			
Air system exhibits no audible leaks				Boom length indicator functioning			
Tire pressure acceptable and tire not damaged				Boom angle indicator functioning			
Telescoping boom exhibits no damage to structure, wear pads, boom stops, or cylinder				Engine: hydraulic, air, electrical, oil pressure, temperature, and fuel			
Wire rope free of dirt, excess lube, kinks, and wires and spooled correctly				Operational Inspection			
Reeving correct				Correct counterweight for the load			
Wedge sockets and wire rope clips not distorted, cracked, or missing				Main hoist control functioning			
Block not damaged				Auxiliary hoist control functioning			
Ball and hook is free to swivel and rotate				Anti-two block in place and functioning			
Guards are in place				Swing brake			
Outrigger float(s) secured with pad pin				Lights and horns functional			
Cab							
Handrails in place and not damaged							
Operator's manual in vehicle							
Load chart legible and visible to operator							
Hand signal chart visible to workers							
Charged fire extinguisher in place							
Cab glass not cracked and wipers are functional							

Notes:

Date _____

Operator
Crane number

Signature	
Model	

Visual Inspection	Pass	Fail	N/A
Engine fluid level correct (check dip stick or sight glass)			
Hydraulic fluid level correct (check dip stick or sight glass)			
Hydraulic system exhibits no apparent weeping or leaks			
Air system exhibits no audible leaks			
Tire pressure acceptable and tire not damaged			
Telescoping boom exhibits no damage to structure, wear pads, boom stops, or cylinder			
Wire rope free of dirt, excess lube, kinks, and wires and spooled correctly			
Reeving correct			
Wedge sockets and wire rope clips not distorted, cracked, or missing			
Block not damaged			
Ball and hook is free to swivel and rotate			
Guards are in place			
Outrigger float(s) secured with pad pin			
Cab			
Handrails in place and not damaged			
Operator's manual in vehicle			
Load chart legible and visible to operator			
Hand signal chart visible to workers			
Charged fire extinguisher in place			
Cab glass not cracked and wipers are functional			

	Pass	Fail	N/A
Gauges and Indicators			
Load moment indicator operational			
Drum rotation indicator functioning			
Boom length indicator functioning			
Boom angle indicator functioning			
Engine: hydraulic, air, electrical, oil pressure, temperature, and fuel			
Operational Inspection			
Correct counterweight for the load			
Main hoist control functioning			
Auxiliary hoist control functioning			
Anti-two block in place and functioning			
Swing brake			
Lights and horns functional			

Notes:

Date_____

Operator		Signature	
Crane number		Model	

Visual Inspection	Pass	Fail	N/A		Pass	Fail	N/A
Engine fluid level correct (check dip stick or sight glass)				**Gauges and Indicators**			
Hydraulic fluid level correct (check dip stick or sight glass)				Load moment indicator operational			
Hydraulic system exhibits no apparent weeping or leaks				Drum rotation indicator functioning			
Air system exhibits no audible leaks				Boom length indicator functioning			
Tire pressure acceptable and tire not damaged				Boom angle indicator functioning			
Telescoping boom exhibits no damage to structure, wear pads, boom stops, or cylinder				Engine: hydraulic, air, electrical, oil pressure, temperature, and fuel			
Wire rope free of dirt, excess lube, kinks, and wires and spooled correctly				**Operational Inspection**			
Reeving correct				Correct counterweight for the load			
Wedge sockets and wire rope clips not distorted, cracked, or missing				Main hoist control functioning			
Block not damaged				Auxiliary hoist control functioning			
Ball and hook is free to swivel and rotate				Anti-two block in place and functioning			
Guards are in place				Swing brake			
Outrigger float(s) secured with pad pin				Lights and horns functional			
Cab							
Handrails in place and not damaged							
Operator's manual in vehicle							
Load chart legible and visible to operator							
Hand signal chart visible to workers							
Charged fire extinguisher in place							
Cab glass not cracked and wipers are functional							

Notes:

Date_____

Operator
Crane number

Signature	
Model	

Visual Inspection	Pass	Fail	N/A
Engine fluid level correct (check dip stick or sight glass)			
Hydraulic fluid level correct (check dip stick or sight glass)			
Hydraulic system exhibits no apparent weeping or leaks			
Air system exhibits no audible leaks			
Tire pressure acceptable and tire not damaged			
Telescoping boom exhibits no damage to structure, wear pads, boom stops, or cylinder			
Wire rope free of dirt, excess lube, kinks, and wires and spooled correctly			
Reeving correct			
Wedge sockets and wire rope clips not distorted, cracked, or missing			
Block not damaged			
Ball and hook is free to swivel and rotate			
Guards are in place			
Outrigger float(s) secured with pad pin			
Cab			
Handrails in place and not damaged			
Operator's manual in vehicle			
Load chart legible and visible to operator			
Hand signal chart visible to workers			
Charged fire extinguisher in place			
Cab glass not cracked and wipers are functional			

	Pass	Fail	N/A
Gauges and Indicators			
Load moment indicator operational			
Drum rotation indicator functioning			
Boom length indicator functioning			
Boom angle indicator functioning			
Engine: hydraulic, air, electrical, oil pressure, temperature, and fuel			
Operational Inspection			
Correct counterweight for the load			
Main hoist control functioning			
Auxiliary hoist control functioning			
Anti-two block in place and functioning			
Swing brake			
Lights and horns functional			

Notes:

Date_____

Operator		Signature	
Crane number		Model	

Visual Inspection	Pass	Fail	N/A		Pass	Fail	N/A
Engine fluid level correct (check dip stick or sight glass)				**Gauges and Indicators**			
Hydraulic fluid level correct (check dip stick or sight glass)				Load moment indicator operational			
Hydraulic system exhibits no apparent weeping or leaks				Drum rotation indicator functioning			
Air system exhibits no audible leaks				Boom length indicator functioning			
Tire pressure acceptable and tire not damaged				Boom angle indicator functioning			
Telescoping boom exhibits no damage to structure, wear pads, boom stops, or cylinder				Engine: hydraulic, air, electrical, oil pressure, temperature, and fuel			
Wire rope free of dirt, excess lube, kinks, and wires and spooled correctly				**Operational Inspection**			
Reeving correct				Correct counterweight for the load			
Wedge sockets and wire rope clips not distorted, cracked, or missing				Main hoist control functioning			
Block not damaged				Auxiliary hoist control functioning			
Ball and hook is free to swivel and rotate				Anti-two block in place and functioning			
Guards are in place				Swing brake			
Outrigger float(s) secured with pad pin				Lights and horns functional			
Cab							
Handrails in place and not damaged							
Operator's manual in vehicle							
Load chart legible and visible to operator							
Hand signal chart visible to workers							
Charged fire extinguisher in place							
Cab glass not cracked and wipers are functional							

Notes:

Date_____

Operator
Crane number

Signature	
Model	

Visual Inspection	Pass	Fail	N/A
Engine fluid level correct (check dip stick or sight glass)			
Hydraulic fluid level correct (check dip stick or sight glass)			
Hydraulic system exhibits no apparent weeping or leaks			
Air system exhibits no audible leaks			
Tire pressure acceptable and tire not damaged			
Telescoping boom exhibits no damage to structure, wear pads, boom stops, or cylinder			
Wire rope free of dirt, excess lube, kinks, and wires and spooled correctly			
Reeving correct			
Wedge sockets and wire rope clips not distorted, cracked, or missing			
Block not damaged			
Ball and hook is free to swivel and rotate			
Guards are in place			
Outrigger float(s) secured with pad pin			
Cab			
Handrails in place and not damaged			
Operator's manual in vehicle			
Load chart legible and visible to operator			
Hand signal chart visible to workers			
Charged fire extinguisher in place			
Cab glass not cracked and wipers are functional			

	Pass	Fail	N/A
Gauges and Indicators			
Load moment indicator operational			
Drum rotation indicator functioning			
Boom length indicator functioning			
Boom angle indicator functioning			
Engine: hydraulic, air, electrical, oil pressure, temperature, and fuel			
Operational Inspection			
Correct counterweight for the load			
Main hoist control functioning			
Auxiliary hoist control functioning			
Anti-two block in place and functioning			
Swing brake			
Lights and horns functional			

Notes:

Date _____

Operator		Signature	
Crane number		Model	

Visual Inspection	Pass	Fail	N/A		Pass	Fail	N/A
Engine fluid level correct (check dip stick or sight glass)				**Gauges and Indicators**			
Hydraulic fluid level correct (check dip stick or sight glass)				Load moment indicator operational			
Hydraulic system exhibits no apparent weeping or leaks				Drum rotation indicator functioning			
Air system exhibits no audible leaks				Boom length indicator functioning			
Tire pressure acceptable and tire not damaged				Boom angle indicator functioning			
Telescoping boom exhibits no damage to structure, wear pads, boom stops, or cylinder				Engine: hydraulic, air, electrical, oil pressure, temperature, and fuel			
Wire rope free of dirt, excess lube, kinks, and wires and spooled correctly				**Operational Inspection**			
Reeving correct				Correct counterweight for the load			
Wedge sockets and wire rope clips not distorted, cracked, or missing				Main hoist control functioning			
Block not damaged				Auxiliary hoist control functioning			
Ball and hook is free to swivel and rotate				Anti-two block in place and functioning			
Guards are in place				Swing brake			
Outrigger float(s) secured with pad pin				Lights and horns functional			
Cab							
Handrails in place and not damaged							
Operator's manual in vehicle							
Load chart legible and visible to operator							
Hand signal chart visible to workers							
Charged fire extinguisher in place							
Cab glass not cracked and wipers are functional							

Notes:

Date_____

Operator
Crane number

Signature	
Model	

Visual Inspection	Pass	Fail	N/A
Engine fluid level correct (check dip stick or sight glass)			
Hydraulic fluid level correct (check dip stick or sight glass)			
Hydraulic system exhibits no apparent weeping or leaks			
Air system exhibits no audible leaks			
Tire pressure acceptable and tire not damaged			
Telescoping boom exhibits no damage to structure, wear pads, boom stops, or cylinder			
Wire rope free of dirt, excess lube, kinks, and wires and spooled correctly			
Reeving correct			
Wedge sockets and wire rope clips not distorted, cracked, or missing			
Block not damaged			
Ball and hook is free to swivel and rotate			
Guards are in place			
Outrigger float(s) secured with pad pin			
Cab			
Handrails in place and not damaged			
Operator's manual in vehicle			
Load chart legible and visible to operator			
Hand signal chart visible to workers			
Charged fire extinguisher in place			
Cab glass not cracked and wipers are functional			

	Pass	Fail	N/A
Gauges and Indicators			
Load moment indicator operational			
Drum rotation indicator functioning			
Boom length indicator functioning			
Boom angle indicator functioning			
Engine: hydraulic, air, electrical, oil pressure, temperature, and fuel			
Operational Inspection			
Correct counterweight for the load			
Main hoist control functioning			
Auxiliary hoist control functioning			
Anti-two block in place and functioning			
Swing brake			
Lights and horns functional			

Notes:

Date _____

Operator		Signature	
Crane number		Model	

Visual Inspection	Pass	Fail	N/A		Pass	Fail	N/A
Engine fluid level correct (check dip stick or sight glass)				**Gauges and Indicators**			
Hydraulic fluid level correct (check dip stick or sight glass)				Load moment indicator operational			
Hydraulic system exhibits no apparent weeping or leaks				Drum rotation indicator functioning			
Air system exhibits no audible leaks				Boom length indicator functioning			
Tire pressure acceptable and tire not damaged				Boom angle indicator functioning			
Telescoping boom exhibits no damage to structure, wear pads, boom stops, or cylinder				Engine: hydraulic, air, electrical, oil pressure, temperature, and fuel			
Wire rope free of dirt, excess lube, kinks, and wires and spooled correctly				**Operational Inspection**			
Reeving correct				Correct counterweight for the load			
Wedge sockets and wire rope clips not distorted, cracked, or missing				Main hoist control functioning			
Block not damaged				Auxiliary hoist control functioning			
Ball and hook is free to swivel and rotate				Anti-two block in place and functioning			
Guards are in place				Swing brake			
Outrigger float(s) secured with pad pin				Lights and horns functional			
Cab							
Handrails in place and not damaged							
Operator's manual in vehicle							
Load chart legible and visible to operator							
Hand signal chart visible to workers							
Charged fire extinguisher in place							
Cab glass not cracked and wipers are functional							

Notes:

Date _____

Operator		Signature	
Crane number		Model	

Visual Inspection	Pass	Fail	N/A
Engine fluid level correct (check dip stick or sight glass)			
Hydraulic fluid level correct (check dip stick or sight glass)			
Hydraulic system exhibits no apparent weeping or leaks			
Air system exhibits no audible leaks			
Tire pressure acceptable and tire not damaged			
Telescoping boom exhibits no damage to structure, wear pads, boom stops, or cylinder			
Wire rope free of dirt, excess lube, kinks, and wires and spooled correctly			
Reeving correct			
Wedge sockets and wire rope clips not distorted, cracked, or missing			
Block not damaged			
Ball and hook is free to swivel and rotate			
Guards are in place			
Outrigger float(s) secured with pad pin			
Cab			
Handrails in place and not damaged			
Operator's manual in vehicle			
Load chart legible and visible to operator			
Hand signal chart visible to workers			
Charged fire extinguisher in place			
Cab glass not cracked and wipers are functional			

	Pass	Fail	N/A
Gauges and Indicators			
Load moment indicator operational			
Drum rotation indicator functioning			
Boom length indicator functioning			
Boom angle indicator functioning			
Engine: hydraulic, air, electrical, oil pressure, temperature, and fuel			
Operational Inspection			
Correct counterweight for the load			
Main hoist control functioning			
Auxiliary hoist control functioning			
Anti-two block in place and functioning			
Swing brake			
Lights and horns functional			

Notes:

Date_____

Operator	
Crane number	

Signature	
Model	

Visual Inspection	Pass	Fail	N/A
Engine fluid level correct (check dip stick or sight glass)			
Hydraulic fluid level correct (check dip stick or sight glass)			
Hydraulic system exhibits no apparent weeping or leaks			
Air system exhibits no audible leaks			
Tire pressure acceptable and tire not damaged			
Telescoping boom exhibits no damage to structure, wear pads, boom stops, or cylinder			
Wire rope free of dirt, excess lube, kinks, and wires and spooled correctly			
Reeving correct			
Wedge sockets and wire rope clips not distorted, cracked, or missing			
Block not damaged			
Ball and hook is free to swivel and rotate			
Guards are in place			
Outrigger float(s) secured with pad pin			
Cab			
Handrails in place and not damaged			
Operator's manual in vehicle			
Load chart legible and visible to operator			
Hand signal chart visible to workers			
Charged fire extinguisher in place			
Cab glass not cracked and wipers are functional			

	Pass	Fail	N/A
Gauges and Indicators			
Load moment indicator operational			
Drum rotation indicator functioning			
Boom length indicator functioning			
Boom angle indicator functioning			
Engine: hydraulic, air, electrical, oil pressure, temperature, and fuel			
Operational Inspection			
Correct counterweight for the load			
Main hoist control functioning			
Auxiliary hoist control functioning			
Anti-two block in place and functioning			
Swing brake			
Lights and horns functional			

Notes:

Date _____

Operator		Signature	
Crane number		Model	

Visual Inspection	Pass	Fail	N/A
Engine fluid level correct (check dip stick or sight glass)			
Hydraulic fluid level correct (check dip stick or sight glass)			
Hydraulic system exhibits no apparent weeping or leaks			
Air system exhibits no audible leaks			
Tire pressure acceptable and tire not damaged			
Telescoping boom exhibits no damage to structure, wear pads, boom stops, or cylinder			
Wire rope free of dirt, excess lube, kinks, and wires and spooled correctly			
Reeving correct			
Wedge sockets and wire rope clips not distorted, cracked, or missing			
Block not damaged			
Ball and hook is free to swivel and rotate			
Guards are in place			
Outrigger float(s) secured with pad pin			
Cab			
Handrails in place and not damaged			
Operator's manual in vehicle			
Load chart legible and visible to operator			
Hand signal chart visible to workers			
Charged fire extinguisher in place			
Cab glass not cracked and wipers are functional			

	Pass	Fail	N/A
Gauges and Indicators			
Load moment indicator operational			
Drum rotation indicator functioning			
Boom length indicator functioning			
Boom angle indicator functioning			
Engine: hydraulic, air, electrical, oil pressure, temperature, and fuel			
Operational Inspection			
Correct counterweight for the load			
Main hoist control functioning			
Auxiliary hoist control functioning			
Anti-two block in place and functioning			
Swing brake			
Lights and horns functional			

Notes:

Date_____

Operator		Signature	
Crane number		Model	

Visual Inspection	Pass	Fail	N/A		Pass	Fail	N/A
Engine fluid level correct (check dip stick or sight glass)				**Gauges and Indicators**			
Hydraulic fluid level correct (check dip stick or sight glass)				Load moment indicator operational			
Hydraulic system exhibits no apparent weeping or leaks				Drum rotation indicator functioning			
Air system exhibits no audible leaks				Boom length indicator functioning			
Tire pressure acceptable and tire not damaged				Boom angle indicator functioning			
Telescoping boom exhibits no damage to structure, wear pads, boom stops, or cylinder				Engine: hydraulic, air, electrical, oil pressure, temperature, and fuel			
Wire rope free of dirt, excess lube, kinks, and wires and spooled correctly				**Operational Inspection**			
Reeving correct				Correct counterweight for the load			
Wedge sockets and wire rope clips not distorted, cracked, or missing				Main hoist control functioning			
Block not damaged				Auxiliary hoist control functioning			
Ball and hook is free to swivel and rotate				Anti-two block in place and functioning			
Guards are in place				Swing brake			
Outrigger float(s) secured with pad pin				Lights and horns functional			
Cab							
Handrails in place and not damaged							
Operator's manual in vehicle							
Load chart legible and visible to operator							
Hand signal chart visible to workers							
Charged fire extinguisher in place							
Cab glass not cracked and wipers are functional							

Notes:

Date_____

Operator		Signature	
Crane number		Model	

Visual Inspection	Pass	Fail	N/A
Engine fluid level correct (check dip stick or sight glass)			
Hydraulic fluid level correct (check dip stick or sight glass)			
Hydraulic system exhibits no apparent weeping or leaks			
Air system exhibits no audible leaks			
Tire pressure acceptable and tire not damaged			
Telescoping boom exhibits no damage to structure, wear pads, boom stops, or cylinder			
Wire rope free of dirt, excess lube, kinks, and wires and spooled correctly			
Reeving correct			
Wedge sockets and wire rope clips not distorted, cracked, or missing			
Block not damaged			
Ball and hook is free to swivel and rotate			
Guards are in place			
Outrigger float(s) secured with pad pin			
Cab			
Handrails in place and not damaged			
Operator's manual in vehicle			
Load chart legible and visible to operator			
Hand signal chart visible to workers			
Charged fire extinguisher in place			
Cab glass not cracked and wipers are functional			

	Pass	Fail	N/A
Gauges and Indicators			
Load moment indicator operational			
Drum rotation indicator functioning			
Boom length indicator functioning			
Boom angle indicator functioning			
Engine: hydraulic, air, electrical, oil pressure, temperature, and fuel			
Operational Inspection			
Correct counterweight for the load			
Main hoist control functioning			
Auxiliary hoist control functioning			
Anti-two block in place and functioning			
Swing brake			
Lights and horns functional			

Notes:

Made in the USA
Coppell, TX
13 September 2021